FORBIDDEN SCIENCE

From Ancient Technologies
to Free Energy

Edited by J. Douglas Kenyon

Bear & Company
Rochester, Vermont

Bear & Company
One Park Street
Rochester, Vermont 05767
www.BearandCompanyBooks.com

Bear & Company is a division of Inner Traditions International

Library of Congress Cataloging-in-Publication Data
Forbidden science : from ancient technologies to free energy / edited by J. Douglas
Kenyon.
     p. cm.
   Summary: "Reveals the cutting edge of new science on myriad fronts and shows
how established science disallows inquiry based on conjecture—even when it
produces verifiable results"—Provided by publisher.
   Includes bibliographical references.
   ISBN 978-1-59143-082-7
   1. Science—Miscellanea. I. Kenyon, J. Douglas.
   Q173.F64 2008
   500—dc22
                                                                    2007042672

Printed and bound in the United States by Versa Press, Inc.

10   9   8

Text design by Rachel Goldenberg, with layout by Priscilla Baker
This book was typeset in Sabon, with Avant Garde used as a display typeface

All pictures are reprinted from *Atlantis Rising* magazine unless otherwise noted.

Contents

Part THREE • Challenges to Traditional Physics

Part FOUR • Spiritual Science

Part FIVE • Astronomy Farther Out

Part SIX • Medicine of Another Kind

Part NINE • Other Dimensions

Part TEN • The Future

Introduction

J. Douglas Kenyon

This book explores some of the less traveled, even darkened, corridors beneath the shining edifice of academic science. In these pages you will find evidence that, no matter what the mandarins of the establishment would claim, the truth is not nearly so exemplary, or easily dismissed. In these pages you will learn of many controversial notions supposedly debunked by conventional argument, if, in fact, they have been discussed at all. But from the true function of the Great Pyramid and the megaliths at Nabta Playa to Immanuel Velikovsky's astronomical insights, from zero-point energy and cold fusion to Rupert Sheldrake's research into telepathy and ESP, we think you will see that the facts are something quite different from what you may have been led to believe. And, if, in the end, you ask yourself why such material is excluded from consensus thought—indeed, why a discussion of it has been virtually "forbidden"—then you are asking yourself the same difficult questions as are the authors of this book.

For those who speak a particular language, it is easy to extract meaning from its expressions, but for those who have not learned the tongue, it all seems to be just so much noise. I can still remember how as a child, upon hearing incomprehensible talk in another language, I thought I could fool others into thinking I understood by making up my own gibberish. Alas, the tactic didn't work and I remember only blank stares for my pains. Eventually I learned that one man's eloquence is another man's gibberish. The difference is understanding.

A couple of MIT grad students have recently taken the theme of confusion of tongues to new heights in the real academic world. The students, Reuters reported, successfully passed off a bunch of computer-generated gibberish as an academic paper. Using a program that they had written to generate fake research complete with nonsensical text, charts, and diagrams, they submitted two of their papers to the World Multi-conference on Systemics, Cybernetics, and Informatics (WMSCI), scheduled in Orlando, Florida. To their surprise, one of the papers—"Rooter: A Methodology for the Typical Unification of Access Points and Redundancy"—was accepted for presentation.

The episode reminded me of a personal experience: as a college freshman, many years ago (at a school that shall here go nameless), I criticized the quality of writing in the school's poetry journal. Someone told me that if I was so smart, I should try submitting something myself. I said I would. Forthwith I produced what I considered to be a really bad poem, but of the type the journal seemed to like, and sent it in. To my amazement, my entry was not only printed, but was also featured on the front cover. My case was made.

The intention here is not just to suggest that many of the so-called arbiters of knowledge occupying the seats of authority in today's citadels of scientific authority may actually be faking it, but also to point out that their criticism of many in the alternative science community—many of whom really *do* know something—should, perhaps, be taken with a grain of salt.

Over the years I have noticed that many who think they actually know the rationales of alternative science often respond with talking points that skirt the real issues and focus mostly on trivialities. The so-called skeptics of the Committee for Scientific Investigation of Claims of the Paranormal (CSICOP) and like organizations seem incapable of understanding the language they are offering to translate. Or as John Anthony West is fond of saying, "They just don't get it." All that is thus demonstrated is their own ignorance.

Another dimension of the problem is in the business world, where some look at publications like this and see what they take to be simple niche targeting. Mystified by the actual content, these observers then conclude that a similar successful result can be achieved by simply compiling a collection of gobbledygook and labeling it with the appropriate buzzwords. The fundamental coherence of the scientific studies and perspectives presented here seems to have been missed, and they think they can match it with gibberish. They may be surprised to discover what you already know: our goal is to make sense, not money, never mind that we are talking to an ever-growing audience.

Although the so-called mainstream media attempt to convince everyone that the subjects covered in this volume should be placed entirely under the heading of fringe science, it so happens that what the scientific establishment consigns to the fringe the vast majority of the public puts much closer to the center of its concerns. A recent Gallup Poll, in fact, reports "about three in four Americans profess at least one paranormal belief." The most popular is extrasensory perception (ESP). In this area, at least, the pronouncements of the scientific elite regarding what we may or may not believe can be countermanded by the evidence of our own senses. The old sales pitch "Who are you going to believe? Me or your lying eyes?" may be failing again. Certainly many of us have personally witnessed many things that orthodox science cannot explain.

The Gallup Poll is not the only recent indicator of weakness in the scientific edifice. According to the Louis Finkelstein Institute for Social and Religious Research, more than 60 percent of doctors reject the Darwinian belief that "humans evolved naturally with no supernatural involvement." According to Drs. Michael A. Glueck and Robert J. Cihak, writing for *Jewish World Review* online, doctors know too much about the actual working of the body to be impressed by the simplicities offered in the Darwinian perspective. One example cited is the human eye—an amazing complex that shows all the hallmarks of a designed system that the standard evolutionary model cannot begin to explain.

The polling data that should probably be the most troubling to official science, however, comes from Minnesota's Health Partners Research Foundation. According to a report published in the British journal *Nature,* one in three U.S. scientists, in an anonymous survey, admitted to breaking—in the last three years—rules designed to ensure the honesty of their work. The misbehaviors, says the *Minneapolis Star Tribune,* range from claiming credit for someone else's work to changing study results due to pressure from a sponsor. "Our findings suggest," say the authors, "that U.S. scientists engage in a range of behaviors extending far beyond falsification, fabrication, and plagiarism that can damage the integrity of science."

Is the agenda of such disingenuous science larger than science itself? Could political power of some kind be the aim?

During the recent federal court trial in Dover, Pennsylvania, which commanded the attention of the world, over whether "intelligent design" (ID) may be taught in schools, it was fascinating to watch classic strategies and tactics for achieving political advantage acted out. Once again we were reminded that, indeed, there is nothing new under the sun.

Take, for example, the use of the word *evolution.* There is nothing in the current ID theory that denies the reality of evolution. In fact, much of the serious work on ID has to do with determining just what evolution itself may require in order to function. One needs a chicken to have an egg and an egg to get a chicken, so, clearly, evolution may occasionally need some help (i.e., an intelligent designer), but that does not mean *evolution* (i.e., progressive change) is not happening. On the contrary, it is clear that such change is a fact, and serious proponents of intelligent design have no quarrel with it.

To us—notwithstanding accusations that ID is antievolution and antiscience—it seems that it may offer more of an enlightened middle way between false choices. Heretofore, we have been told we must select either biblical creationism or evolution. But what is challenged in the current debate is not evolution; it is "Darwinism"—the idea that evolution could occur with only random and material forces at work, no intelligence involved. Ironically, those who "believe" the latter are in actuality staking out a metaphysical position and holding to it by faith without proof (i.e., adhering to a dogma) and are advancing a virtual religion of their own while claiming to reject the authority of any religion.

The cult of Darwinism, it seems to us, has usurped the role of the priesthood that it has ostensibly overthrown, suggesting that it and only it can provide the answers the world is seeking. And all the while, its adherents feign an air of injured innocence when its integrity is questioned and its authority challenged.

Though the inner workings of minds at the upper echelons of the Darwinian religion may be difficult for nonbelievers to fathom, we can still study their influence at lower, less informed levels of the ecclesiastical hierarchy and make some useful observations. For instance, when called to defend the "sacred" cause of "science"—perceived as threatened by ID's growing influence—much of the secular press has obediently rushed to the ramparts. The shrill, even hysterical, denunciation of ID as nothing more than a front for biblical fundamentalist creationism and the prophesying of doom—a virtual return to the Dark Ages—reveals, however, more than the ignorance of the accusers. Indeed, the cries of alarm over the imminent "death of science," we suspect, reflect diminishing certainty and growing anxiety over the authority of the entire Darwinian position. In such a state, argument based on merit alone is far too threatening and must be abandoned.

The authors of this book—among many others—have observed a deep disconnect in the orthodox truth-detection mechanisms of our society. The difficulty isn't so much due to a conspiracy as to a schism tearing apart the very soul of civilization. The result, it can be argued, has been a host of problems: alienation, wars, environmental collapse, and so on. One symptom of the disorder is the elevation of the unworthy into positions of authority from which they—in an endless effort to preserve their own advantages—manipulate the levers of power. And where there is opportunity to corrupt, there is, it seems, no shortage of willing corrupters. The condition is widespread and intolerable. But, hopefully, in the current conflict over intelligent design, we are witnessing one of those extraordinary moments when the system, in the interest of its own equilibrium, is entering into some much needed self-correction.

If that indeed is what is happening, we may witness some intense resistance.

So, what else is new?

In the *Chicago Sun Times* review of the recently released Hollywood fantasy *Eragon,* film critic Miriam Di Nunzio complains that she just doesn't understand why the black magician Durza "cannot simply wave his hands and retrieve" the missing blue stone sought by the evil king and his minions. And later she questions the logic of a story in which the villain is surprised to discover that there are forces hostile to the king who should have been exterminated. "Why Durza can't magically divine this is beyond me," is her exasperated comment. The answer to Di Nunzio's complaint, though, may have been provided in the film itself when Brom, played by Jeremy Irons, says, "Magic has rules" ("rules" being another word for "laws").

We mention the argument over *Eragon* not because we consider it a particularly noteworthy movie, but rather to illustrate a point. Di Nunzio seems to be among those who assign anything paranormal to a realm that has no basis in reality. According to this way of thinking, any story of magic is, by definition, fiction, where the only rules are those created by the author. In other words, once you decide to tell a tall tale, what's the point in letting something like logic slow you down?

This simplistic way of thinking dominates in today's media mainstream and elsewhere. Ironically, it is from this quarter that the epithet "supernatural" seems most frequently deployed. The orthodox assumption is that we have the familiar natural world that is obedient to the basic laws of physics as we understand them; anything else must be "super"-natural—unbound by natural laws—and, of course, not real. By this way of thinking, anything we don't understand becomes "supernatural"—that is, "strictly imaginary." The most obvious battle line here is between those, such as religious fundamentalists, who believe the supernatural *exists* (their "God," who created natural law, doesn't have to obey it if he doesn't want to) and those militant secularists who believe it *does not exist* and that our present "scientific" grasp of reality is not to be questioned. Only a few, it seems, are left to argue that ultimately reason itself depends upon the notion that *order,* whether understood or not, is supreme, and that the appearance of any unexplained phenomena eventually tells us more about the limits of our "understanding" than about the limits of the natural order.

Strangely, though, many of those self-anointed custodians of our current collection of principles said to comprise natural law (a.k.a. "the paradigm keepers," themselves fundamentalists from a different church) seem unwilling to or incapable of perceiving possibilities that might exist outside the box of our current understanding. These "high priests" of consensus reductionist science are fond of classifying anyone who does not accept the limitations that they have imposed upon reality as believers in the "supernatural" or worse. To put it another way, they classify those who see things differently as ignorant and superstitious, if not dabblers in the black arts.

However, as Arthur C. Clarke famously observed, "any sufficiently advanced technology is indistinguishable from magic." It is clear that many of our existing technologies can produce effects that even our immediate ancestors would have considered magic, so why does our hubris prevent us from seeing that things we cannot currently comprehend might not be so incomprehensible if we but knew a little more? Indeed, isn't it reasonable to suggest that many of our most cherished assumptions—the rules that we now believe govern reality—may be in need of expansion and revision, and that even our distant ancestors may have understood things that we have now forgotten but someday, hopefully, will understand again?

In times like these, it is worth remembering—if we may be permitted to mix two old refrains—"now we see as through a glass darkly" but "further along we'll understand more."

When self-styled experts in the revealed wisdom of the current order become irate, the real question they need to answer is what exactly has them so upset? If they are so certain of the plausibility of their case, what possible concern can they have over our "rantings"? Methinks that they—having doubts about their own position, which they would rather not discuss—protest too much.

In a recent online debate over the reality of the afterlife, the defender of the skeptical position declared to his opponent, "I don't know that it [the afterlife] does *not* exist, but you don't know that it *does*." We have seen similar comments on bumper stickers. The general implication seems to be that anyone claiming knowledge that exceeds that of the "skeptic" cannot possibly be sincere and, thus, must be lying, with ulterior motives to boot. This kind of rhetoric from the debunker hit squads has become standard fare in many fields—from the afterlife to intelligent design, from zero-point energy to antigravity—and it is pursued with an emotional fervor that is hard to ignore. Just what, we wondered, should be inferred from such behavior?

Could it be that the institutional mystique that evokes such awe from the media and much of the public is nothing so much as an elaborate subterfuge intended to disguise the weakness and the blindness of these entrenched elitists—something which, like the emperor's new clothes, even a child could perceive? We'll leave the conspiracy angles to others, but it seems apparent that, at least on a subconscious level, much of the posturing, if not the bullying, betrays what is at best a deep insecurity about the actual validity of their claims. The very speed with which some of the more outspoken take offense at any suggestion that the basic paradigm of materialist reductionist science can be questioned betrays, we suspect, deep-seated doubts in their own ability to discern, much less discuss, the truth.

Let's put it another way. Suppose the so-called debunkers and their brethren were colorblind and realized their disadvantage over those who could actually see color. Their need to level the playing field—to deny the reality of color and to demand that those who could perceive objects and relationships that they could not (such as traffic lights being red, not green) be labeled as charlatans, or worse—might be understandable, but it would not be defensible. The tactic would, of course, lose out as long as these color skeptics were out of power, but if their tribe were to take over and then prop up their weaknesses with the force of law, wouldn't all who could see rainbows become outlaws?

So far, we remain free to promote awareness of the many hues that adorn our world, some of which may be hard to see unless properly pointed out. But books like this one could be very threatening to those who perceive the world as strictly black and white—or, at best, as shades of gray. Let us hope they are not permitted to enforce their insecurities.

On the other hand, for those who may find the inherent dangers of our time a bit overwhelming, there is still plenty of reason to take heart. The discoveries and knowledge cited in these pages, to say nothing of the heroism, could show all of us the pathway to liberation that we have been seeking. Indeed, if the way ahead looks perilous, it is worth remembering that it has never been otherwise. As a wise man once said, "All change occurs through dramatic episode."

PART ONE

STACKING THE DECK

1 Debunking the Debunkers

Do the So-called Skeptics Have a
Secret Agenda?

David Lewis

If you believe in the paranormal, or life after death, you better watch out. The cops might show up at your door—the *Psi Cops*, members of the Committee for Scientific Investigation of Claims of the Paranormal, or CSICOP. These skeptics spend a lot of time and energy debunking anything scientifically offbeat or extrasensory in nature. They work tirelessly, trying to enforce the unenforceable "law" that says no phenomenon can exist beyond the notion of a purely physically based reality. Phonetically, their acronym suits them, *psi* being the nickname scientists use for extrasensory phenomena—hence, "Psi Cops." And they have their hands full these days, what with all those bestsellers about near-death experiences, angels, and lost civilizations.

Crime has really gotten out of control.

Books about the universe having conscious origins, the new consciousness-based physics, has CSICOP chairman Paul Kurtz in a dither too. At a recent skeptics conference in New York City, he stated that postmodernists (the new physics movement) deny absolute scientific knowledge is possible, the result being an "erosion of the cognitive process, which may *undermine democracy*" [emphasis added]. Sounds awfully serious.

Recognizing the paranormal, according to Kurtz, questions the prevailing scientific worldview, and that's just too scary for his Psi Cops to think about. At a CSICOP meeting featuring Harvard's John Mack, a renowned psychiatrist (now deceased) who researched claims of alien abductions, the debate took on an inquisitional tone.

To Mack's surprise, a skeptic announced she had infiltrated his pool of abductees, good Psi Cop that she was, the idea being that Mack's acceptance of her charade diminished his credibility. Mack took a lot of heat that day, and it was surely embarrassing. But he questioned the Psi Cops' vehemence and dogma, reminding them that other cultures have always known about "other realities, other beings, other dimensions . . . that can cross over into our own world." Doing so, Mack irked the skeptics even more. Paul Kurtz later lamented, "If we allow Mack's suggestion, then we have to allow for angels and past lives. Where does it all stop?"

Crime in the streets, no doubt.

Reincarnation, astrology, and spirituality have no place in the debunkers' worldview, likewise homeopathy and Linus Pauling, and the list goes on and on. Even conspiracy theories about the assassination of JFK frazzle a debunker's sensibilities. As champions of Francis Bacon's scientific method, a system of drawing conclusions

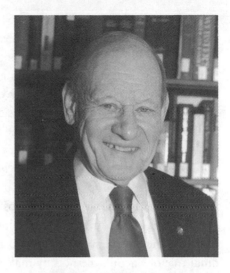

Fig. 1.1. Prominent psi debunker and CSICOP chairman Paul Kurtz.

from observable fact rather than from assumption, these "skeptics" present themselves as priests of pure science. But it turns out they practice what they condemn most, a "belief system" known as scientific materialism, the doctrine to which Bacon's method devolves when scientists trade free thought and inquiry for the dogma of absolute materialism.

A scientific materialist believes matter is the only truth, that everything in the universe, including consciousness, can be explained in terms of physical laws—no transcendent cause, no purpose, no meaning to life.

In short, our thoughts, feelings, inspirations, identity—the universe itself—are merely highly evolved chemical reactions. The soul, of course,

Fig. 1.2. John Mack (1929–2004), psychiatrist and professor at Harvard Medical School, was a leading authority on alleged alien abduction experiences.

does not exist for the scientific materialist, nor does any awareness beyond the brain, nor anything vaguely spiritual in nature, acceptance of which they disparagingly refer to as "superstition." This cynicism is extended to any area that challenges the prevailing academic view, including theories of advanced lost civilizations, alternative medicine, and the paranormal. The theory advanced by Boston University's Robert Schoch and author John Anthony West, for instance, that the Sphinx may be far older than previously thought, as evidenced by water erosion, meets with a hail of criticism, not necessarily on scientific grounds but because the implications challenge the prevailing assumptions about prehistory. From consciousness-based reality to theories about advanced lost civilizations, paradigms that force a reevaluation of our origins, they claim it's all hogwash. All evidence to the contrary, they deem such theories to be fraud or flawed, violating the cardinal rule of Bacon's method by making a priori assumptions, all the while claiming the highest standard of intellectual purity.

How did Wayne and Garth of *Saturday Night Live* fame put it? "We are not worthy . . . we are not worthy."

To give their movement pizzazz, the Psi Cops enlisted the likes of Carl Sagan, ex-magician turned debunker James Randi, comedian Steve Allen, and an assortment of academics who share their nihilistic beliefs. Their purpose is to convince the "superstitious" that belief in anything but nuts-and-bolts materialism is hokum, thereby saving us, and democracy, from our better instincts. Their skepticism is absolute and, of course, unproved, yet advanced as fact by much of the academic and scientific communities. This absolute skepticism is the hidden premise behind every position the debunkers take, never mind nagging problems like "Where did all that Big Bang energy come from in the first place?"

The problem, says John Beloff, a Scottish psychologist at the University of Edinburgh, lies in their "skeptical position." To his credit, Kurtz published a paper Beloff wrote for CSICOP's journal, the *Skeptical Inquirer*. In that Beloff is well known in the field of parapsychology, this was an admittedly unusual occurrence for that publication. In his paper, Beloff discusses the skeptical position, revealing that a priori beliefs exclude the validity of phenomena inconsistent with known, or assumed, physical laws—that means the Psi Cops put the fix in from the start. Beloff summarizes their skeptical position, stating: "Parapsychological findings [to Kurtz] may . . . in due course be taken at face value but always with the tacit understanding that they can eventually be reconciled with a physicalist worldview." Beloff goes on, saying, "Hence, he [Kurtz] specifically rejects the term *paranormal* if this is taken to imply any kind of spiritual, mental, or idealistic dimensions." Dr. Beloff also tells us that Kurtz's position of "absolute skepticism" is by no means unusual. Rather, it is widely shared in the academic and scientific communities. But it's running into trouble.

Ironically, advancements in the field of medicine have precipitated a body of evidence suggesting, perhaps proving, that consciousness exists after death. Similar testimony from hundreds of people, originally compiled by Dr. Raymond Moody in his book *Life After Life*, testifies to a transcendent-beyond-the-body reality. Case after case of clinically dead people coming back to life in hospital emergency rooms challenge the skeptics to apply their materialist views in new and creative ways. TV programs dealing with near-death experiences trot out skeptics who condescendingly relegate the profound spiritual episodes of resuscitated patients to the realm of neurotransmitters, hallucinations, and fraud, certain that the brain alone is the source of consciousness. Rather few in number, these skeptics surface in the media frequently. Presenting the obligatory opposing viewpoint, they ignore evidence that contradicts their assumptions, such as clinically dead patients recalling conversations in the waiting room, after they, the patients, *expired* in the emergency room and then came back to life.

Dr. Kenneth Ring's *Life at Death: A Scientific Investigation of the Near-Death Experience* points toward a paradigm shift leading to recognition of the primary role of consciousness in reality. His conclusions strike at the heart of scientific materialism and absolute skepticism, pulling the rug out from under the Psi Cops. "The world of modern physics and the spiritual world seem to reflect a *single* reality" [his emphasis], Ring states. He also admonishes that material science has its limits and that the pursuit of absolute knowledge lies in the realm of religion, philosophy, and spirituality. His position isn't new. Mystics, intellectuals, and influential scientists have made the same point. Albert Einstein put it poetically, saying, "The most beautiful thing we can experience is the mysterious. It is the source of all true art and science. He to whom this emotion is a stranger, is as good as dead: his eyes closed. . . . To know what is impenetrable to us really exists, manifesting itself as the highest wisdom and the most radiant beauty which our dull faculties can comprehend only in their most primitive forms, this knowledge, this feeling, is at the center of true religiousness."

We need to rise to the challenge of being more like Einstein, seeking out the mystery of life, in spite of the nagging voices of scientific materialism, which may simply reflect our own collective mistrust of intuition and inspiration. At the same time, we should not ignore what the skeptics have to offer, a rigorous application of critical thinking in areas prone to superstition and charlatanism. The scientific method has and will serve us well if properly understood. It got us out of the Dark Ages and into the space age, cured polio, and so on (although science recognizes that discovery often results from accident). But a troubling marriage sometimes aligns scientific materialism with those who attack anyone embracing nontraditional systems. As the cult of absolute materialism finds its way into our lives, schools, and the courtroom, we run the risk of diminishing personal liberty and free thought, *real* threats to democracy. In the name of science, debunkers, skeptics, and "experts" suddenly don hats of authority, seemingly with the imprimatur of the scientific community.

The January/February 1995 issue of the *Skeptical Inquirer* features such an "expert." In that issue, Joseph Szimhart writes disparagingly about James Redfield's best-selling novel, *The Celestine Prophecy,* which Szimhart evaluates, for some reason, as if it were a work of nonfiction. Had Szimhart simply not liked the book or its content there would be little to say. Had his background been accurately represented by the *Skeptical Inquirer,* there would again be little to say. But Szimhart impugns Redfield's character without any supporting evidence, suggesting his only motive for telling his story is money; he also assaults religious and mystical traditions and their exponents, including the Maharishi Mahesh Yogi, Baird Spalding, Guy Ballard, and Carlos Castaneda. As in Redfield's case, he implies money is their only motive. He calls Nicholas Notavitch's account of Jesus Christ's journey to India "bogus," maligning a tradition that has existed for two thousand years, though only more recently in the West. He then describes the widely popular *A Course in Miracles* as a "reactionary . . . dictatorial tome."

Lighten up, Joe.

But Szimhart's intellectual prejudice is not his only problem. His background as a self-styled "deprogrammer" presents a deeper set of concerns. He has reacted to New Age belief systems maniacally, forcibly detaining and intimidating people involved with what some scholars call "New Religious Movements." Charged with kidnapping in an Idaho case, Szimhart narrowly escaped conviction though his accomplices did not. Subsequently, ex-coworkers have denounced his fanatical methods. His is a profession that, according to a Syracuse University study, may induce post-traumatic stress disorder in the people he coerces and detains, doing far more damage than if they had been left alone.

His diary, seized by the authorities, reveals his motive for working with kidnappers: money. His article in the *Skeptical Inquirer* reveals yet another motivation: his peculiar antipathy for anything resembling "awaken[ing] to this inner reality, or gnosis" (his own words). His intolerance and unsavory dealings somehow earn him the title "Specialist in Controversial New Religions" in the footnote to his article. The *Skeptical Inquirer*'s editor, one would think, might apply his skepticism more evenly.

Fortunately, few skeptics share Szimhart's tactics or fanaticism. He is no scientist, and genuine skeptics may wonder why his work appeared in the *Skeptical Inquirer* in the first place. Moreover, many scientists, some calling themselves skeptics, approach claims of paranormal phenomena with genuine objectivity. Others actively investigate the mysterious, the offbeat, and the transcendental. Theories and evidence of consciousness-based reality have captured the attention of notable scientists and professionals, like Harvard's John Mack, as mentioned, and physicist/Nobel laureate Brian Josephson, who writes about "The Next Grand Union, Physics and Spirituality."

Skeptics come unglued, of course, when distinguished professionals cross over into the forbidden zone of consciousness exploration. John Mack had the audacity to study claims of alien abduction, bizarre accounts of people claiming to have been kidnapped by extraterrestrials and experimented on while under telepathic control, accounts that suggest a merging of subconscious and physical realities. After exhausting all other explanations, Mack took the accounts, recalled under hypnosis, at face value, theorizing that *reality* must be more than it seems. As a result, his tenure at Harvard came under review and he was denounced by some of his peers, while other professionals saluted his courage.

Brian Josephson stunned his colleagues when he turned to consciousness exploration after having discovered a magical quantum property now called the Josephson effect (the phenomenon of current flow across two weakly coupled superconductors separated by an insulating barrier) at the University of Cambridge at the tender age of twenty-two. He then received a tenured position at

Fig. 1.3. Physicist and Nobel laureate Brian Josephson, director of the Mind-Matter Unification Project at the Cavendish Laboratory in Cambridge, England.

Cambridge's legendary Cavendish Laboratory. That was in 1972. He won a Nobel Prize a year later. Subsequently, he renounced the world of orthodoxy for the pursuit of mystical understanding. The scientific community considered Josephson a genius, until he crossed into the "forbidden zone." But his inclinations showed up early on, when as a graduate student he revealed his appreciation for invisible realities. His discovery of the Josephson effect came as a result of his having theorized that electron "tunnels" might pass through insulating barriers in superconducting circuits the same way ghosts pass through walls in the movies.

Based on his reading of quantum mechanics, the inner workings of the universe, he guessed that the current in such a circuit could actually flow in both directions at once, creating a kind of standing wave that would be especially sensitive to magnetic and electrical influences. Bell Laboratories validated Josephson's theories, adding to his already growing reputation as an innovator and a prodigy. In a recent issue of *Scientific American,* he says quantum mechanics allows for "synchronicities" that "produce the appearance of psychic phenomena." Decoded, that means consciousness-based physical reality as opposed to the other way around. When he lectures at the Cavendish Laboratory, his views are well received, he says. In the same article, Josephson suggests scientists can improve their abilities through the practice of meditation.

We might say that hard-and-fast skeptics lack Josephson's subtlety of mind. This is not to say that all skeptics reject out of hand what Josephson represents. To the contrary, some pursue the truth in earnest, wherever it leads, such as Dr. Michael Epstein, a chemist and the vice president of a skeptics group. Epstein commented in a news release for the Society for Scientific Exploration that "[d]ebunkers often call themselves skeptics. However, a real skeptic is one who is willing to look critically at all of the evidence for extraordinary claims, and that's what SSE is here to do."

The society, a group of scientists and academics, met in Huntington Beach, California, recently. Topics discussed ranged from near-death experiences (NDEs) to evidence for cases of reincarnation, enough to frazzle any Psi Cop. Other topics dealt with biological responses that may predict earthquakes, the effect of the moon on human behavior, artificial structures on Mars, the age of the Sphinx, sacred sites and sacred science, acoustical properties of ancient ceremonial sites, archaeoastronomy, alternative energy, inertia loss in spacecraft, and telepathy and psychokinesis. Society members do not necessarily subscribe to the positions presented. Rather, they apply a scientific standard that neither rejects nor accepts theories out of hand. Professor Lawrence Frederick, for instance, secretary of the society and former secretary of the American Astronomical Society, rejects the methodology that has been used to gather evidence for artificial structures on Mars but does not rule out the theory altogether. Frederick speaks candidly regarding the monuments on Mars theory, saying, "I can't show that it isn't true, but it sounds goofy." Without a double-blind test, he says, using other locations on Mars against which to compare the geometry of the supposed artificial structures, a scientific conclusion cannot be drawn.

Yet Frederick and society members investigate with an open mind, which others will not. They champion free

Fig. 1.4. Challenging today's debunker is not unlike fighting a dragon—but there is a lot more smoke to deal with.

inquiry into a wide range of theories and claims, no matter how strange. In their voices, one hears a blend of fascination and skepticism, perhaps the ideal mixture of scientific rigor and human wonder. Speaking about one member, who shall remain nameless, Frederick describes him as an "informative and lovely person," tenured at a major polytechnic institute. The reason he shall remain nameless: while he sides with the Psi Cops on most issues, he's convinced the Loch Ness monster really exists. . . . Honest. His position, of course, presents a serious problem. It makes you wonder.

What will become of democracy?

2 "Voodoo Science" on Trial

Challenging the Establishment's Kangaroo Court of Alternative Science

Eugene Mallove, Ph.D.

*V*oodoo *Science,* published in 2000, may well be regarded by future historians of science as a dying ember from the funeral pyre of late-twentieth-century-establishment physics, which hurtles toward a supposed "theory of everything" while being blissfully ignorant of profound cracks in its very foundations. But author Robert L. Park, a physics professor at the University of Maryland, is now riding high. For some years he has been the darling of editors seeking crisp commentary from the chief representative of the American Physical Society (APS), a position he has held since 1982.

Whether railing against manned space flight, antiballistic missile defense, alternative health care, ESP research, UFO investigation, or his favorite whipping topic, cold fusion, Robert Park can be found in top mudslinging form on the op-ed pages of the *New York Times* and the *Washington Post,* among others. His politicized weekly "What's New" Internet science column (www.opa.org/WN) is remarkable in that it is tolerated at all by the APS, especially since Park, with insufferable chutzpah, ends each column with a fake disclaimer: "Opinions are the author's and are not necessarily shared by the APS, but they should be." That's pure Park, who hopes that his audience will come to see the world through the filter of the scientific certainties that he and many of his arrogant physics colleagues claim to possess.

Dr. Park has now compiled his "wisdom" in a short volume in which he claims to have discovered a new kind of science—"Voodoo Science." His definition of voodoo science is encapsulated in the subtitle, "The Road from Foolishness to Fraud." There is a progression from "honest error" that evolves "from self-delusion to fraud," he says. He further elaborates the definition: "The line between foolishness and fraud is thin. Because it is not always easy to tell when that line is crossed, I use the term 'voodoo science' to cover them all: pathological science, junk science, pseudoscience, and fraudulent science."

He says that he "discovered" voodoo science in the course of his pub-

Fig. 2.1. Author and physics professor Robert Park, cold fusion adversary and self-appointed gatekeeper of conventional scientific thought.

lic relations work for the APS: he "kept bumping up against scientific ideas and claims that are totally, indisputably, extravagantly wrong." He is that certain, three adverbs worth, that many of the things he calls "voodoo science" cannot be right. More often than not, he draws his conclusions from fundamental theory that is supposedly sacrosanct. Therein lies the fundamental failure of Park and so many of his colleagues in the physics establishment: they have abandoned what little curiosity about scientific experiments that they may have had at the beginning of their scientific careers. They attack data from experiments that at first glance appear to be in conflict with theory, about which they have concluded one of two things: 1) The theory can't possibly need fundamental modification, which might allow the phenomenon to occur, or 2) It is inconceivable that existing theory can be applied to allow the phenomenon. It takes a special kind of arrogance to conclude affirmatively on both those points, particularly when both experimental data and theory for an anomalous phenomenon trend strongly against the doubters, the study of cold fusion being a prime example.

Park thinks he knows what he and the physics establishment are doing, but he does not. He writes ". . . no matter how plausible a theory seems to be, experiment gets the final word." But for Park, theory rules which experiments he will even look at. Revealing complete ignorance of the bloody battles over paradigm shifts in science (of the very kind he is obstructing!), Park claims, "When better information is available, science textbooks are rewritten with hardly a backward glance." Baloney!

In *Voodoo Science,* Park dismisses cold fusion at its very first mention, referring to it as "the discredited 'cold fusion' claim made several years earlier by Stanley Pons and Martin Fleischmann."* He says that a "dwindling

---

*On March 23, 1989, in Salt Lake City, scientists Stanley Pons and Martin Fleischmann announced their discovery of "cold fusion"—nuclear reactions occurring near room temperature that could generate electrical power.

Fig. 2.2. Stanley Pons (far left) and Martin Fleischmann (far right), the electrochemists who discovered cold fusion in 1989. (Photograph courtesy of "Infinite Energy")

band of believers" continue to gather each year "at some swank international resort" in an attempt to "resuscitate" cold fusion. He asks, "Why does this little band so fervently believe in something the rest of the scientific community rejected as fantasy years earlier?" He speculates later, "Perhaps many scientists found in cold fusion relief from boredom."

Park works himself up about cold fusion throughout the book and tells us what he really thinks of cold fusion: "On June 6, 1989, just seventy-five days after the Salt Lake City announcement, cold fusion had clearly crossed the line from foolishness to fraud." He states that Fleischmann and Pons "exaggerated or fabricated their evidence." (He only speculates whether cold fusion researcher Dr. James Patterson of Clean Energy Technologies Inc. may have "crossed the line from foolishness to fraud.")

Park has not troubled himself to study the very data that he demanded many years ago as proof of cold fusion, specifically the helium-4 nuclear ash data, even after this data made it into the peer-reviewed literature. On June 14, 1989, in the *Chronicle of Higher Education*, Park opined, "The most frustrating aspect of this controversy is that it could have been settled weeks ago. If fusion occurs at the level that the two scientists claim, then helium, the end product of fusion, must be present in the used palladium cathodes." "You don't have to worry about the heat if there is no helium" was his statement to me in the spring of 1991, recorded in my book *Fire from Ice*. Apart from his gross error of ignoring the helium that might be in the cover gas coming from surface reactions (such cold fusion helium had been detected in 1991 and later), it is notable that Park has never mentioned any of the published literature on helium in cold fusion experiments.

Since at least 1991, Park has been informed by fellow APS scientists, such as Dr. Scott Chubb, about helium-4 detection in cathodes and in the gas streams of cold fusion experiments. These independent experiments have been published in the United States and Japan in peer-reviewed journals. There is no doubt that Park knows this. But *Voodoo Science* contains no mention of this data, an egregious fraud by Park on journalists and the general public.

Thus, on the issue of cold fusion, Park himself has traveled, in his lexicon, from foolishness to fraud. Not troubling himself with inconvenient facts, such as experimental evidence of robust character that supports cold fusion, he states preposterously: "Ten years after the announcement of cold fusion, results are no more persuasive than those in the first weeks." He rewrites cold fusion history with ludicrous bloopers designed to entertain: "How, I wondered, could Pons and Fleischmann have been working on their cold fusion idea for five years, as they claimed, without going to the library to find out what was already known about hydrogen in metals?" Electrochemist Martin Fleischmann, Fellow of the Royal Society, not knowing a lot about hydrogen in metals? A bit much to suggest, even for an unethical obfuscator like Park. Park is the one who should have gone to the library. He would have discovered that leading cold fusion scientists like Fleischmann and Bockris wrote the textbooks about hydrogen in metals. Fleischmann's outstanding research in this area earned him the Fellowship in the Royal Society, arguably the world's most prestigious scientific society. In other contexts Park claims allegiance to established theory and the expertise of leading authorities; in this case, he does not even realize who the authorities are.

If Park doesn't get his information about cold fusion from technical papers, the normal approach in science, from where does he get it? Apparently he is briefed by fact-resistant critic Dr. Douglas Morrison, of CERN (European Organization for Nuclear Research), who has attended the international cold fusion conferences where he asks mostly obtuse questions, proving that he, like Park, has not read the cold fusion literature. Morrison has "kept an eye on cold fusion for the rest of us," as Park puts it. The result of all this is to have Morrison, the prime purveyor of the "pathological science" theory of cold fusion, passing misinformation to Park, who then jazzes it up with snide remarks suited to the Washington beltway crowd.

While Morrison is the only skeptic to actually publish a paper that attempts to come to grips with quantitative issues of cold fusion calorimetry and electrochemistry, every paragraph in his paper includes an elementary mistake. A few examples: he subtracts the same factor twice. He claims that Fleischmann and Pons used "a complicated non-linear regression analysis" method, which they did not use. He recommends another method instead— the one, in fact, they did use. He confuses power (watts) with energy (joules). He claims that hydrogen escaping from a 0.0044 mole palladium hydride might produce 144 watts of power and 1.1 million joules of energy, whereas the textbooks say the maximum power from this would be 0.005 watt, and a simple calculation shows that the most energy it could produce is 650 joules. This is the "expert" Park relies upon for news of cold fusion!

But Park well knows the propaganda value of turning a serious subject into a joke. In his account of the early days of cold fusion he observes, "Cold Fusion was becoming a joke. In Washington that is usually fatal."

After assaulting the main body of cold fusion research, Park singles out for attack Dr. Randell Mills of BlackLight Power Inc. (see *Infinite Energy* 17:21–35). He says that Mills did not offer "any experimental evidence" for his claims of excess energy caused by catalytic hydrino formation. Park does not discuss the multiple channels of experimental and astrophysical data that Mills has cited to defend his theory. Park covers up the serious, positive results that the NASA Lewis Research Center published in its official report on the Mills replication. But Park, at his core, argues mainly from theory: "But those who bet on hydrinos are betting against the most firmly established and successful laws of physics." Mr. Certain asks rhetorically, "What are the odds that Randall [sic] Mills is right? To within a very high degree of accuracy, the odds are zero."

Though I expected Park to bash scientific anomalies, I was unprepared to discover the depths of his ignorance about space flight and its future. Commenting on his early 1990s testimony before Congress in support of unmanned space missions, he recalls, "I wanted to explain why the era of human space exploration had ended twenty-five years earlier and was unlikely ever to come back." No future for human presence in space? Is Park for real? He ends his myopic refrain with inept poetry bearing an absurd message, "America's astronauts have been left stranded in low-earth orbit, like passengers waiting beside an abandoned stretch of track for a train that will never come, bypassed by the advance of science."

Amateur astronaut Park then offers an amazing blooper: "If there was gold in low earth orbit, it would not pay to get it." Astonishing! When he makes this and other claims he apparently does not understand such elementary concepts as the small propulsion energy cost of de-orbiting with rockets and aerobraking. In the emerging era of commercial space transportation, this Park faux pas will be remembered as a late-twentieth-century howler, on par with statements by astronomer Simon Newcomb earlier in the century that heavier-than-air flight was likely to remain impossible.

In Park's crusade against manned space flight, he even goes after astronaut hero John Glenn: "Both Ham [a chimpanzee aboard an early U.S. space flight] and Glenn would end up in Washington: Glenn in the U.S. Senate, Ham in the national zoo. Ham died a short time later without ever returning to space."

He attacks "messianic engineer" Robert Zubrin, who has put forth concrete, well-researched proposals for cost-saving space missions in his book *The Case for Mars*. Park says that Zubrin started "his own cult—the Mars Society." Park mocks the aspirations that led people like Dr. Robert Goddard and so many others this century to work toward the manned exploration of space: "Zubrin had learned his lessons well. The focus is on the dream. His followers feel their feet crunching into the sands of Mars, while the most daunting technical challenges are swept aside with simplistic solutions."

On the book jacket Park singles out "magnet therapy" in addition to cold fusion as the epitome of "foolish and fraudulent scientific claims." In the only "experiment" that he actually decided to personally conduct to test any of his opinions, he launched a misguided effort to disprove the alleged therapeutic effectiveness of magnets in contact with the human body. He bought some athletic magnets from a local store, then stuck one on a steel file cabinet. He then inserted sheets of paper under the magnet, and found that at ten sheets the magnet fell off. He exults, "Credit cards and pregnant women are safe! The field of these magnets would hardly extend through the skin, much less penetrate muscles." Park had merely found the point at which static friction (caused by the magnetic force) is insufficient to hold the magnet against the force of gravity. On this basis he concludes that the magnetic field would not penetrate into skin! This is completely wrong, as sophomore physics students at MIT, and presumably at the University of Maryland, would know. Park gets a grade of F on that one. "Not that it would make any difference if it did penetrate," he says. Park always has some a priori theoretical insight about why something "can't be." This public relations agent for the American Physical Society needs a refresher course in Science 101.

Given Park's incompetent assessment of cold fusion and his failures in elementary scientific methodology, we cannot expect a useful appraisal of other controversial areas, such as whether or not there are loopholes in or extensions to classical thermodynamics, whether low-level electromagnetic fields can affect biological systems, the "memory of water," or the scientific foundations of alternative medicine. Regardless of their individual merits, Park gives all of these questions the same brush-off he applies to cold fusion.

It is not that one might never find areas of agreement with Park. For example, some of the charlatan-like antics of Dennis Lee, of Better World Technologies, which Park chronicles, are appalling and have nothing to do with the serious scientific investigation of anomalous energy phenomena. And Park states that "there is now overwhelming scientific evidence that we ourselves can affect Earth's climate." Some scientists would agree with that; I don't happen to. I side with those atmospheric scientists who believe that computer models do not yet come close to an adequate representation of all the factors that affect climate.

On the other side, Park is rather forgiving about such things as government spending for Tokamak hot fusion, which is widely regarded as a financially wasteful research boondoggle even by those who have nothing to do with cold fusion. He says absolutely nothing about the ill-fated Superconducting Supercollider (SSC), which was begun and then canceled before it could waste even more taxpayer money. We do not hear of the scandalous recent cost overrun at the ICF (Inertial Confinement Fusion) weapons simulation laser fusion device, which was led by a physicist who was not even honest about his academic credentials. To Park, this waste is apparently "all in the family"—the kind of money that the white-collar welfare, government-funded physics community can be forgiven for wasting.

It is tempting to speculate that Park may be suffering from a psychological problem known as projection, or possibly cognitive dissonance. At some level, this confused man with all his years of schooling must realize that he is out of his element in evaluating the cold fusion evidence. He doesn't really know whether the evidence is good or not. Obviously he has not studied it except superficially, yet he has gone far out on a limb in attacking it—he can't bring himself to turn back. Among other problems, admitting he had been very wrong would call into question his many other judgments, from manned space travel to magnet therapy. He expected that cold fusion would go away years ago but it hasn't, so he creates the myth that the cold fusion field consists of "followers who see what they expect to see." In truth, it is Park who is seeing what he wants to see—lack of evidence where there is evidence! The following grand assessment by Park of "voodoo scientists" pertains most properly to him: "While it never pays to underestimate the human capacity for self-deception, they must at some point begin to realize that things are not behaving as they had supposed." It will be cosmic justice for this profoundly foolish, mean-spirited flak for the physics establishment when, in the light of scientific advance, the bigotry and lies he has turned against others expose him for what he is.

3 The Establishment Strikes Back

Never Mind the Evidence, The Learning
Channel Wants to Eradicate the Looming
Atlantis Heresy

Frank Joseph

Our times are witnessing an unprecedented general interest in Atlantis. The release in the summer of 2001 of Walt Disney Studios' feature film *Atlantis, the Lost Empire* was a reflection of that worldwide fascination. Serious and competent investigators employing the latest research technology are making paradigm-breaking finds, from the waters around the Bahamas and Cuba to the Bolivian Alto Plano and the mid-Atlantic Ocean. The very appearance and descriptive name of the magazine in which this article first appeared, *Atlantis Rising*, tells it all—Atlantis is rising in the consciousness of millions as never before in modern history. It is only natural then, that this resurgence would provoke reaction from conventional scholars who regard any mention of any "Antediluvian World" as the worst heresy. Doubtless, the growing popularity of "the A word" has caused them deep frustration, after so many years of concerted attempts at debunking it in front of audiences from the classroom to television-land. Mortified but unbowed, the defenders of Ivory Tower archaeology are repeatedly airing a new "special" over The Learning Channel entitled "Uncovering Atlantis."

I want to state clearly at the outset that even the most fanatical Atlantologist does not oppose the public discussion of contrary views, but would, moreover, enjoy confronting them as healthy challenges. I, like my colleagues who research Atlantis, enjoy any opportunity to contrast our facts and ideas with the shopworn dogma of uniformitarians. Honest differences of opinion and lively debate are catalysts of discovery. The most strident opposition is welcome, as long as it is in the spirit of genuine scientific curiosity. But when that opposition sets out to deliberately vilify a targeted viewpoint and its adherents with blatant untruths, it must be exposed and condemned.

For its first five minutes, "Uncovering Atlantis" entices viewers with suggestions of some historical credibility for the sunken kingdom. They are told that it was described by one of the greatest minds of the Western world, the

Greek philosopher Plato. Cultural comparisons between the ancient Near East and pre-Columbian America, the narrator continues, imply a vanished, mid-ocean source common to them both. But this deceptively sympathetic attitude soon deteriorates into imperious denial with the appearance of Dr. Kenneth Feder, who teaches at Central Connecticut State University.

Shocked to learn that four out of five of his new students entertain the possibility that Atlantis might have actually existed, he has taken an extraordinary precaution against such intolerable open-mindedness. Before getting into the basics of Archaeology 101, Feder regularly subjects his captive audiences to an anti-Atlantis indoctrination class, reproduced in the TV program. The very notion of a lost civilization is ridiculed point for point in a one-sided presentation that leaves no room for debate. "That there is no reference to Atlantis before Plato, if such a place really existed," emotes Feder, "is absolutely stunning." The disembodied voice of the narrator adds, "Atlantis was forgotten after Plato's death for more than two thousand years. The first person to mention or perhaps invent this word was Plato."

In truth, variations of the history of Atlantis were known literally around the world among dozens or perhaps hundreds of disparate societies centuries before Plato. In many of these indigenous traditions, Atlantis is self-evidently referred to and culturally inflected, such as Aztlan, a volcanic "White Island" in the "Sunrise Sea" from which the ancestors of the Aztecs arrived on the eastern shores of Mexico. Another "White Island" was described in the great Indian epics the *Mahabharata* and *Puranas* as Attala, the mountainous homeland of a powerful and highly civilized race located in the "Western Sea" on the other side of the world from India. The *Vishnu Purana* located Attala "on the seventh zone," which corresponds to 24–28 degrees latitude, on a line with the Canary Islands. There, the Guanche inhabitants spoke of Atara, while Atemet was the dwelling place of an Egyptian goddess responsible for a world-flood.

In North American Cherokee tradition, Atali is the place from which their ancestors dispersed throughout the world immediately after a catastrophic deluge. Named after the lost homeland of the Maya's earliest ancestors, Atitlan is a lake region in the Solola Department, the central highlands of southwestern Guatemala, where Quiche-Maya civilization reached its florescence. In Euskara, the language of the Basque people, Atlaintika is the name of a drowned kingdom from which their forebears arrived in the Bay of Biscay. Atlatonan was the "Daughter of Atlaloc," a blue-robed virgin ritually drowned to venerate the Aztec rain god. Her fate and philological resemblance to Atlantis, literally, "Daughter of Atlas," is too remarkable for coincidence.

Contrary to the narrator of "Uncovering Atlantis," the sunken civilization was not forgotten for two thousand years after Plato. During classical times, it was the subject of lively discussion among the leading thinkers of the Greco-Roman world including Aristotle, Strabo, Poseidonus, Krantor, Proclus, Plutarch, and Diodorus Siculus, most of whom, incidentally, believed in the historical reality of Atlantis. So did the seventeenth century's greatest scholars, Athanasius Kircher in Germany and Sweden's Olof Rudbeck. The narrator asks, "[T]he archaeologists' verdict [on Atlantis]?" Various archaeologists caustically respond, one after the other on camera, "Garbage!" "Insidious!" "Misleading!" "Preposterous!" "Fantasy!" Yet these exclamations more accurately describe The Learning Channel's own pseudo-documentary. Certainly, the narration is "preposterous" when describing its self-styled debunker: "Ken Feder has become an expert in recognizing the kind of evidence that will prove whether or not Atlantis was the source of all civilization." The professor is then shown sifting Native American pottery, among which he finds pieces of early colonial ware. Since he was unable to find evidence of a similar Atlantean intrusion at his little informal surface dig, he concludes that there is no evidence for ancient visitors from "the lost continent" in rural Connecticut. Sadly, this kind of paltry argument falls somewhat short of proving "whether or not Atlantis was the source of all civilization."

Such pathetic attempts at debunking the lost empire are supported by a narration asserting that any apparent resemblances between pyramids of the Old and New Worlds are entirely coincidental and unrelated. As Feder pontificates, "The pyramids can definitely be ruled out as any connection to Atlantis." He denies that Mesoamerican pyramids bear any comparison with Sumerian ziggurats and especially the early Egyptian step pyramids. But the Third Dynasty pyramid built by Pharaoh Zoser at Saqqara compares in numerous details with the Mayan pyramid at Palenque, in Yucatán. Beyond their obvious outward resemblance as step pyramids, both contain subterranean chambers with descending corridors. The Palenque pyramid entombed the body of a Mayan monarch known as Pacal. The funeral practices and beliefs that accompanied him were remarkably similar to those known in the Old World. The eight degrees this king's soul was expected to pass on its way through the Underworld were virtually identical in Mesoamerican and Egyptian belief systems. In the latter, the third degree was overseen by a crocodile, the *sibak; cipak* is Nahuatl for the Aztecs' funeral alligator-boat.

The human soul was identically conceived in Egyptian and Mesoamerican thought. Temple art across the Nile Valley portrayed the *ba* as a human-headed bird, often soaring through a hole in a tomb. A Mayan temple relief at Izapa similarly depicts a bird with the head of a man flying out of a burial cave. Wall paintings at Palenque are strikingly reminiscent of the Egyptian technique. Mayan figures, like the Egyptian, are arrayed in rows, while the feet and heads of most of the nobles are portrayed in flat profile. In funeral finery scarcely different from a pharaoh's, Pacal's sarcophagus and his earrings were decorated with hieroglyphs, and he wore a beaded necklace of precious stones cut in the forms of flowers and fruits that could have almost passed for Nile workmanship. Mirroring the pharaohs, Pacal was even fitted with a death mask. The foot of his sarcophagus, like Pharaoh Zoser's, was made so that the entire enclosure could be stood upright on its base. Pacal also wore a false chin-beard, as did all the pharaohs.

The American surveyor Hugh Harleston Jr. discovered in 1974 that Palenque's Temple of the Inscriptions did not fit standard units of 1.059 meters in the Maya's *hunab* system of measurement. He found, instead, that the structure was perfectly laid out in Egyptian "royal cubits." It contained a large room, 23 feet high with a floor space of 13 by 29 feet, and was roofed by stone beams similar to those in the King's Chamber of Egypt's Great Pyramid. Alberto Ruz Lhuillier, a conservative Mexican archaeologist who discovered Pacal's resting place, admitted that this chamber closely resembled numerous details of its Egyptian counterpart. Around Pacal's elaborately carved stone sarcophagus were placed jade statuettes of the solar deity Kinich-Ahau, "Lord of the Eye of the Sun." They were no different from the *ushabti*, or "answerers," small statues made of faience included in royal Egyptian burials. Significantly, the Egyptian solar god, Horus, was identically known as "Lord of the Eye of the Sun" and revered as the deified embodiment of kingship.

Jade was the most important ceremonial stone in Mesoamerica, because its color symbolized the Atlantic waters across which the Maya's culture-bearing ancestors arrived on the shores of Yucatán. In the Aztec version of the Flood, one among this ancestral company was the princess Chalchuitl, with whom jade was personally identified. Pacal's jade statuettes were, in fact, known as *chalchuitls*.

Worse than denying obvious cultural correspondences between the Old and New Worlds was an attempt by the producers of "Uncovering Atlantis" to characterize anyone who entertains the possibility of a historical Atlantis as a potential mass murderer. While running stock footage of Adolf Hitler and his followers, the narrator says, "Prominent Nazis believed the master race originated in Atlantis. One of the most passionate believers was Heinrich Himmler, head of the SS. Himmler directed Germany's scientists to seek for descendants of the Atlantean super race in places ranging from the Andes to Tibet. They scrutinized the physical features of the natives in search of any shred of evidence to support Himmler's notion that his Aryan ancestors, the Atlanteans, once lived there. These claims of an ancestral heritage from Atlantis fed the Nazis' belief in the supremacy of the Aryan master race."

Talk about "fantasy"! After more than twenty years of research on the subject, I have been unable to identify even one of the "prominent Nazis who believed the master race originated in Atlantis." Nowhere throughout the vast literature on the Third Reich does the word "Atlantis" appear, save rarely and incidentally. It is not mentioned in *Mein Kampf,* nor in any of Hitler's hundreds of speeches. In his voluminous *Table Talk,* he cites it once only in passing, during an after-dinner discussion about prehistoric legends. Himmler neither knew nor cared about anything Atlantean; his focus, as any biography of the man reveals, was limited exclusively to Germany. Alfred Rosenberg, the most prominent Nazi philosopher, says nothing about Atlantis in his major opus, *The Myth of the 20th Century.* The Nazis have been accused of many things, but at least they were guiltless of trying to justify "the supremacy of the Aryan master race" by demonstrating Atlantean origins.

Building on this perverse phantom of Nazi Atlantologists, Feder explains, "When we come to something like the lost continent of Atlantis, we are better off knowing that civilizations developed more or less independently, just so nobody can say 'some people are better than others, some are smarter than others' because we know what happens down the line when we believe that. So, I'm not going to tell you that belief in Atlantis is necessarily the first step toward genocide or holocaust. But what I'm telling you is that we are on a very slippery slope when we believe in fantasies, and that those fantasies lead us down to places where we really don't want to go."

In other words, when we begin to question prevailing doctrine about Atlantis, "we are on a very slippery slope" toward genocidal racism. This unspeakable calumny against people who dare to disagree with an official position can have been generated only by deep fear on the part of some conventional scholars. They sense that a growing body of irresistibly persuasive evidence presented by modern Atlantologists is jeopardizing their career investment in fossilized notions of the human past. It is more than absurd to say that anyone is using Atlantis to prove "some people are better than others, some are smarter than others." Quite the contrary; investigators eagerly seek out the folk traditions of native peoples in many parts of the world as vitally important evidence substantiating and illuminating the Atlantean drama. These are the same traditions uniformatarian scholars refuse to take into account, dismissing them with a patronizing smile as "myths." They presume to know better than indigenous peoples about their own prehistory. Who are the real "racists" here?

The unsavory by-products of academia occur in some people, emotionally unequipped for objectivity, as narrow-minded arrogance and self-contained ignorance. "Uncovering Atlantis" may not disclose anything about the lost civilization, but it does say something about mean-spirited men who do not shrink from calumniating others to preserve their imperiled dogma. It is precisely this intolerant doctrine, however, that is at last beginning to sink into popular disrepute as Atlantis is rising ever stronger in the minds of men and women everywhere.

4 Inquisition—The Trial of Immanuel Velikovsky

Velikovsky's Battle to Get His Seminal Book,
Worlds in Collision, Published

Peter Bros

ack in the forties Immanuel Velikovsky, a Russian-born, credentialed scholar and a language expert, came upon an ancient manuscript that led him to believe that the plagues mentioned in the Bible had actually occurred in history. Buried beneath ancient accounts of reality, he found what he thought might be the source of the biblical plagues, the appearance of a huge comet that, as depicted in ancient Sumerian seals, did battle with the Earth as it passed by in the skies. Velikovsky concluded that the comet was actually Venus, which entered the solar system late—some thousands of years ago—and, as it passed nearby, had knocked the Earth and Mars out of their existing orbits. Eventually the intruder, Velikovsky said, took up its position in a near-curricular orbit between Mercury and Earth.

The scholar's original research was published as a book called *Worlds in Collision*. Velikovsky was very much aware that his conclusions contradicted Newton's "celestial mechanics." If all the planets were in place when the solar system formed, then there was no way that an additional planet could have been added, and certainly not within the last five to ten thousand years. Such a view went beyond mere supposition, theory, or idea. It had become, indeed, a virtual "fact," one more solid than the concrete that built the universities whose professors promulgated many such unchallengeable "facts."

One of those universities was Harvard, located in that bastion of clear skies, Boston. And while Boston may not have

Fig. 4.1. Controversial scientist and author Immanuel Velikovsky (1895–1979).

31

had very clear skies, even in the forties, really eminent astronomers such as Harvard's own Harlow Shapley are—when their minds are unclouded by the "realities" and "facts" that form the foundation for many of our most enduring theories—sometimes capable of seeing the light.

Velikovsky, excited by his historical finds, sought out Shapley simply because the professor was, by far, the most notable astronomer of the time.

Shapley, who had an aversion to reading other people's work, agreed to consider Velikovsky's ideas if a third party he respected would bring it to his attention. He agreed that his fellow educator the eminent Harvard philosopher Horace Kallen would do.

Shapley subsequently saved himself considerable reading. When Kallen, in a letter of praise extolling the virtues of Velikovsky's work, let drop the opinion that if Velikovsky should prove correct, the traditional beliefs of astronomy, among other disciplines, would have to be reconsidered, Shapley went ballistic.

"The sensational claims of Dr. Immanuel Velikovsky fail to interest me," he began, "because his conclusions were pretty obviously based on incompetent data," which was to say: because his conclusions opposed the prevailing theory, the results could not be based on fact.

And Shapley was just getting started. If Velikovsky was right in his basic cometery hypothesis, "then the laws of Newton are false. In other words, if Dr. Velikovsky is right, the rest of us are crazy."

We can envision Shapley's mental struggle here—comparing reality with recall. Newton's celestial mechanics sat at the apex of his astronomical hierarchy and provided the template against which all other realities were to be tested. Here comes some doctor—Velikovsky—with nothing but a medical degree to support his claims to scholarship, painting a picture of reality that contradicted the picture Shapley, and indeed the whole world, fervently believed in. Against Shapley's fixed picture of celestial mechanics—a neat and orderly solar system that would continue to move in the same way forever after as it had from eternities before—Velikovsky was painting a picture of the solar system that produced in Shapley a world of outrage.

Shapley expressed his outrage very much like many express their outrage.

First he checked to see who exactly was causing him such outrage, and then he proceeded to launch an all-out assault on the unfortunate offender.

It can be said that most of us want to bring our rage to some kind of resolution, and when that rage is provoked by a conflict between reality and our recollection of that reality, and that conflict seems to be the invention of some individual, then removing that person might seem to go a long way

*Fig. 4.2. Harvard astronomer
Harlow Shapley, Newton
advocate and Velikovsky
adversary.*

toward resolving our problem. Such is the charitable explanation for Shapley's subsequent behavior, but it hardly seems to explain the lengths to which the eminent scientist would soon go.

Before contacting Shapley, Velikovsky had gone to considerable trouble to find a publisher. After expressing interest, Macmillan assigned venerable editor James Putnam to examine the book's possibilities. The exploration involved not only conducting market research, which showed *Worlds in Collision* to be a valuable property, but also contacting scientists in the field for their opinions. This too produced favorable results, with the curator of the Hayden Planetarium observing that Velikovsky's book provided a good opportunity to reexamine "the underpinning of modern science."

Based on these reports, Macmillan signed a full publishing contract with Velikovsky and *Worlds in Collision* was typeset. As part of the prepublication promotion, *Harper's Magazine* had reporter Eric Larrabee prepare a condensation of the book, which was published with the title *The Day the Sun Stood Still.* The *Reader's Digest, Collier's,* even *Paris-Match* picked up versions, and the publicity even made the cover of *Newsweek.*

Shapley responded by writing to Macmillan on Harvard College Observatory stationery noting that the rumor that Macmillan was canceling publication was a great relief. Shapley, the source of the rumor, forged ahead confidently, revealing that "a few scientists with whom I have talked about this matter [are] astonished that the great Macmillan Company . . . would venture into the Black Arts," adding that Velikovsky's output was "the most arrant nonsense of my experience."

The letter was actually an implied threat to boycott Macmillan's lucrative textbook sales, and Putnam wrote back that he couldn't believe the publication of *Worlds in Collision* would affect long-standing views about the excellence of Macmillan's scientific publications. Shapley replied on the same date that the publication would cut him off from the Macmillan Company, that when he had run into Velikovsky in New York by chance, he had "looked around to see if he had a keeper."

Panicked, George Brett, Macmillan's president, wrote Shapley that he would have a panel of independent scholars review the book in advance of publication. The panel endorsed Velikovsky's research as honest and on a subject of scientific, public, and general interest. Brett authorized publication.

This outcome produced in Shapley and his group of Harvard defenders something very much like frenzy. As president of Science Service, Shapley controlled the publication of *Science News Letter* and he began a prepublication blitz in its pages discrediting Macmillan and Velikovsky, even taking out expensive ads in the *New York Times* calling attention to his own attack. He cried to all who would listen that Velikovsky was a crank, his book "the most successful fraud that has been perpetrated on leading American publications," comparing it to the perennially popular Flat Earth tractates (a myth created by evolutionists in the nineteenth century to tar their opponents), complained about the current scientific "age of decadence" and claimed that while he was "a sympathetic friend of the thwarted and demented," Velikovsky's ideas were "rubbish on the level of astrological hocus pocus." He even compared Velikovsky to Senator McCarthy, who spent his time uncovering Soviet spies in the government, a comparison that, considering Velikovsky's Russian roots, seemed rather ironic and strange.

Shapley's frantic efforts went for naught. *Worlds in Collision* was published in April of 1950 and immediately shot to the top of the best-seller lists. While Shapley couldn't threaten to burn Velikovsky at the stake, as the Church had once threatened Galileo and others, science—while claiming to be a fair and open endeavor—soon proved itself to be as dictatorial as the most fanatical of ecclesiastical organizations, as Shapley set about demonstrating its ability to retaliate against apostates.

The first volley was directed at Gordon Atwater, curator at the Hayden Planetarium and chairman of the astronomy department at the American Museum of Natural History. Asked to declare his allegiances regarding Velikovsky, he replied heroically "that science must investigate unorthodox ideas calmly and with an open mind." The response drew Atwater's boss and a colleague to his museum office. The colleague spat in his face and his boss fired him on the spot, forcing him to clear out his office immediately.

Shapley, it turned out, was a member of the museum's board of directors.

With "sheer terror and panic" reigning at the Hayden, *This Week* magazine was pressured to halt publication of an Atwater article. Failing in this effort, Shapley focused on removing potential book reviewers and replacing their work with reviews by his friends, a ploy that worked first at the *Herald Tribune* and then extended throughout the country where the few astronomers who could write contributed reviews calling Velikovsky a crackpot, a liar, and a general threat to the continuation of civilization as we all know it.

After seven weeks of sustained attack, Brett, president at Macmillan, threw in the towel, asking Velikovsky to let Macmillan out of its contract because three-quarters of its business, which derived from textbooks, was in danger.

Velikovsky graciously agreed, allowing the contract to be transferred to the more publicly oriented Doubleday, which not only racked up tremendous profits with it, but also published six additional books authored by Velikovsky. Macmillan, however, which hoped its ordeal was over, found it was just beginning, as the *New York Times* published the facts of the affair in excruciating detail. Undaunted by the negative publicity, Shapley moved in and demanded, as part of Macmillan's penitence, the head of James Putnam, who was dutifully fired after twenty-five years of service to the company.

Reporters, viewing this attack on the freedom of the written word in America and attempting to question Shapley's motives and actions, ironically found themselves called upon to act as the scorpion tail of a scientific community that claimed to have saved humanity from a real catastrophe. The literary critic of the *New Yorker*, for instance, claimed, without a hint of humor, that *Worlds in Collision* was "a pathetic, ominous and superstitious piece of work" whose purpose was to establish a new world order.

Supporters' presentations were called "a mixture of divination, ignorance, haruspices' palaver, and pseudoscientific half-truths," or, more to the point, "plain hokum." The adjectives were colorful. "Not since Captain Hasenpfeffer was reported sailing into New York harbor with a cargo of subways and artesian wells has there been a better candidate for P. T. Barnum's Hall of Fame," moaned the *Christian Science Monitor.* "The most outrageous collection of nonsense since the invention of the printing press" (the *Indianapolis Star*). Velikovsky should be criticized for not including "the Ute legend of cottontail tales of Henny-Penny, Humpty-Dumpty, or even Paul Bunyan and his blue ox Babe" (*The Toronto Globe and Mail*). "A shining example of book and magazine-publishing irresponsibility" (the *Saturday Review of Literature*).

In battling for the preeminence of consensus science, earnest lackeys of the cause frequently call on the writing talents of the not-so-scientifically oriented arts community because they are equally impressed by colorful adjectives and meaningless propositions, many of which can be quite nasty.

The American Association for the Advancement of Science (AAAS) organized a panel discussion on publishing responsibility officered, of course, by Harlow Shapley, at which Macmillan representatives appeared to confess the exact nature of their sins and to beg absolution. AAAS meetings have little to do with science and much to do with ensuring that the penalties for nonconformity with consensus science are very visible to the scientific community.

The reaction to Velikovsky is the standard reaction to apostasy of a science/religion where telescope time, for instance, is denied to credentialed practitioners who fail to confirm the absurd Big Bang concept describing an explosion that created matter. Similarly, peer-reviewed journals refuse to print any criticism of consensus reality, and, in fact, refuse to publish on entire areas that are considered off-limits.

To get around the growing public concern that the entire edifice of empirical science is unsupportable, the latest strategy, devised to avoid the appearance of intolerance resulting from cases like that of Velikovsky, is to simply declare that entire areas that give rise to controversial theories are settled matters—beyond debate. "The distance to the stars," it is said, "has been measured and it's now beyond question. So let's move on to more fruitful discussions like measuring what's going on inside that black hole on the other side of Arcturus."

Velikovsky, in response to the attack, shifted gears, and started playing the game like an empirical scientist. He made several predictions based on his theory, the most notable being that Jupiter emitted radio signals (subsequently found to be correct); he also predicted that Venus would be hot, and he also estimated what turned out to be its actual temperature, all to the derision of the empirical community. Carl Sagan was allowed to write a preemptive article prior to NASA's Venus probe to cover empirical science's behind regarding Venus should Velikovsky be proven correct. But the campaign didn't work, and as the space program generated growing public interest in the sixties, Velikovsky's popularity soared.

Thus, in the early seventies, the AAAS was compelled to hold one of its periodic inquisitions to exorcise itself of evil, arranging an "objective" symposium titled "Velikovsky's Challenge to Science." The symposium, which Velikovsky, in his naïveté, attended, and which was chaired, although not engineered, by Sagan, was a rigged condemnation on everything Velikovsky. It concluded that Velikovsky was a hack and nothing he said was scientific, and therefore his predictions, albeit correct, were not "scientific." The technique continues to serve the ends of consensus science. Witness the current debate over intelligent design.

The High-Technology
Knowledge

PART TWO

THE PAST RECONSIDERED

5 The High Technology of the Ancients

Have We Forgotten Secrets We Once Knew?

Frank Joseph

erek J. de Solla Price got the shock of his life when the true identity of an artifact he happened to be cleaning finally dawned on him. The curious object had lain in Athens' National Museum for half a century after being hauled up from 120 feet below the surface of the eastern Mediterranean Sea, sometime around Easter 1900. It had been found by Elias Stadiatos, a Greek sponge diver working off the coast of Antikythera, a small island near Crete, and part of an ancient Roman freighter that included statues plus other period materials dating the shipwreck to circa 80 BCE.

While examining the object on May 17, 1902, the Greek archaeologist Valerios Stais had noticed a gear wheel embedded in what appeared to be a piece of rock. It was actually a heavily encrusted, badly corroded mechanism in three main parts, comprising dozens of smaller fragments.

The Antikythera Device, as he called it, remained an enigma for the next forty-nine years, until Price, a professor of science history at Yale University, recognized it for what it really was: a mechanical analog computer, an instrument millennia ahead of its time.

"It was like finding a turbo-jet in Tutankamun's [sic] tomb," Price wrote in his June 1959 article for *Scientific American*, "an Ancient Greek Computer." He pointed out that the Antikythera Device uses a differential gear—not reinvented until the mid-1500s—to compute the synodic lunar cycle by subtracting the effects of the sun's movement from effects of sidereal lunar movement, thereby calculating the motions of stars and planets. This function makes the artifact far more advanced than its sixteenth-century differential gear, bringing it into the space age.

The advanced mechanism's function gradually revealed itself after decades of reexamination. When past or future dates were entered via a crank, it calculated the position of the sun, moon, or other astronomical information, such as the location of other planets. The use of differential gears enabled the mechanism to add or subtract angular velocities. Its front dial shows the annual progress of the sun and moon through the zodiac against an Egyptian calendar. The upper rear dial displays a four-year period with associated dials showing the Metonic cycle of 235 synodic months, which approximately equals nineteen solar years. The lower rear dial plots the cycle of a single synodic month, with a secondary dial showing the lunar year of twelve synodic months.

Made of bronze and originally mounted in a wooden frame, the Antikythera Device stands 13 inches high, is 6.75 inches wide but just 3.5 inches thin, and is inscribed with more than two thousand characters. Although most of the text has been deciphered, its complete translation still awaits publication. The complex instrument is displayed in the Bronze Collection of the National Archaeological Museum of Athens, but readers in the United States may see an accurate reconstruction of this ancient analog device at the American Computer Museum in Bozeman, Montana.

The original served as an extremely useful navigation device that enabled the Roman freighter in which it was found to successfully complete transatlantic voyages to America more than fifteen centuries prior to Columbus. And, doubtless, the Antikythera Device was not the first of its kind, but rather the result of prolonged development, stretching back long before it came to rest at the bottom of the Mediterranean Sea in 80 BCE.

The Roman statesman Cicero wrote that the consul Marcellus brought two devices back to Rome from the ransacked city of Syracuse. One device mapped the sky on a sphere and the other predicted the motions of the sun and the moon and the planets. His description seems to match the Antikythera Device, interestingly enough, because Syracuse was the scene of a Roman attack foiled by Archimedes. The Greek mathematical genius arranged a large array of mirrors to reflect sunlight onto the attacking ships, causing them to catch fire. Though modern skeptics believed the account entirely legendary, a group at MIT performed their own tests and concluded that Archimedes' military reflector was feasible after all. The passage of time long ago effaced all traces of his ancient "weapon of mass-destruction," and finds such as the Antikythera Device are extremely rare. But they do suggest that technology in the deep past was far more advanced than mainstream scholars would have us believe.

Among the most surprising, but well-attested, advances actually employed by our ancestors was the submarine, faint memories of which persisted into the Middle Ages, when such a craft was unthinkable, given the Dark Age mentality of the time. *La Vrai Histoire d'Alexandre* [*A True History of Alexander*] is a thirteenth-century French manuscript describing a voyage undertaken by Alexander the Great "in a glass barrel" that carried him from one Greek port to another unnoticed beneath the keels of his fleet of warships in 332 BCE. He was said to have been so satisfied with the submarine's performance that he ordered its production for his navy. If *La Vrai Histoire d'Alexandre* was our only source for such a story, we might be inclined to dismiss it as a medieval fantasy. But Alexander's teacher Aristotle wrote of "submersible chambers" deployed in that same year by Greek sailors during the blockade of Tiros, when the craft stealthily placed underwater obstacles and moored subsurface weapons of some kind.

During Xerxes I's invasion of Europe, a Greek officer, Scyllis, surfaced at night to make his way among all the Persian ships, cutting each vessel loose from its moorings. His submarine used a snorkel, a hollow breathing tube protruding just above the surface of the water. After having set the enemy fleet adrift, Scyllis navigated nine miles back to Cape Artemisium, where he rejoined his fellow Greeks. Similar actions were recounted by some of classical civilization's foremost scholars, including Herodotus (460 BCE) and Pliny the Elder (77 CE).

Around 200 BCE, Chinese chronicles reported the operation of a submarine device carrying one man successfully to the bottom of the sea and back again.

While no ancient submarines have so far been found, physical evidence of a much different kind does survive to prove ancient technology was far in advance of its time. During the late 1990s, the quartz crystal eyes of early dynastic statues were examined by Jay Enoch (School of Optometry, University of California, Berkeley) and Vasudevan Lakshminarayanan (School of Optometry, University of Missouri, St. Louis). They were surprised by the intricacy of anatomical detail found in the artificial eyes of the fourth-dynasty representation of Prince Rahotep and a sculpted scribe from a fifth-dynasty tomb at Saqqara, which the scientists tried to reproduce with the latest optical technology. The ancient Egyptian lenses were found to be of superior quality to the duplications. Enoch and Lakshminarayanan concluded that "because of the performance quality and design complexity, it is highly doubtful that the lenses used to re-create eye structures in ancient Egyptian statues were the first lenses created, despite the fact that they are 4,600 years old."

Their research was complemented by a nearly thirty-year investigation published in 2001. "The earliest actual lenses which I have located," Robert Temple wrote in Australia's *New Dawn* magazine, "are crystal ones dating from the 4th Dynasty of Old Kingdom Egypt, circa 2500 BCE. These are to be found in the Cairo Museum and two are in the Louvre in Paris. But archaeological evidence showing that they must have been around at least seven hundred years earlier has recently been excavated at Abydos, in Upper Egypt. A tomb of a pre-dynastic king there has yielded an ivory knife handle bearing a microscopic carving which could only have been done under considerable magnification (and of course can only be seen with a strong magnifying glass today)."

Temple makes a connection between the mirrors in the Pharos Lighthouse and construction of the Great Pyramid: "The technology for surveying the Great Pyramid existed at least as far back as 3300 BCE, and doubtless earlier than that, since we can hardly presume that the ivory knife handle was the first such object to exist, as it is already highly sophisticated and suggests a long-standing tradition. Thus, we know that magnification technology was in use in Egypt, in 3300 BCE. [The Great Pyramid] is so perfectly oriented to the geographical points of the compass that no one has ever been able to understand how this was done, for the accuracy exceeds any hitherto known technology of ancient Egypt. Then there is the equally famous question of how the extreme accuracy of the construction of the Great Pyramid was possible."

The prominent British Egyptologist Sir Flinders Petrie marveled at "an amount of accuracy [in the Great Pyramid] equal to most modern opticians' straight edges of such a length," and was seconded a century later when Peter Lemesurier observed that its twenty-one acres of polished limestone outer casing "was leveled and honed to the standard of accuracy normal in modern optical work." Egyptian records themselves describe a level of reflective technology mainstream scholars are still reluctant to consider. At sixty feet in height, the 121-ton obelisk at Heliopolis raised for Pharaoh Sesostris I's jubilee in 1942 BCE is the oldest of its kind, and inscribed with a hieroglyphic text describing "13,000 priests chanting before a huge mirror of burnished gold."

In *The Electric Mirror of the Pharos Lighthouse,* author Larry Brian Radka conclusively shows that practical electronics were applied in pharaonic civilization, most obviously at the famous lighthouse. He points out that the amount of flammable fuel that would have been needed to power its beacon never existed in all of Egypt, and imports would not only have been prohibitively expensive, but also exhausted within the first year of its operation. Due to these and other no less cogent considerations, Radka credibly argues that the Pharos Lighthouse featured a carbon-arc lamp in which a blindingly bright light was produced by an electrical spark jumping between the pointed ends of positively and negatively charged rods. He claims that its power source was a bank of liquid, primary cells known as the Lalande Battery, invented (reinvented?) in the nineteenth century by Felix Lalande and Georges Chaperon. The Egyptians possessed all the materials (glass, copper, mercury, and lye) to make its ancient predecessor. As Radka explains, "Several large Lalande cells placed in series and parallel could have supplied enough voltage and current to power the Pharos light for many hours before any of their elements would have needed replacing. This type of battery needs no external source of electricity to revitalize it. After it has discharged, replacing two of its internal ingredients restores the unit to full capacity."

The existence of such a battery is not mere speculation; it is supported by smaller, though fundamentally similar, cell batteries found elsewhere in the ancient Near East, most famously, the so-called Baghdad Battery, discovered by archaeologist Wilhelm Koenig, during 1938, in Stuttgart, Germany. The earthenware jar was fitted with an asphalt stopper pierced by an iron rod, its lower section inside surrounded by a copper cylinder. When filled with common fruit juice, the device generates two volts of electricity. In 1940, Professor Koenig published a scientific paper on the artifact that had originally been found at Khujut Rabu, just outside Baghdad, and dated to 250 BCE, more than two thousand years before the official invention of the electric battery by Alesandro Volta in the early nineteenth century. After the Second World War, Willard F. M. Gray, of the General Electric High Voltage Laboratory in Pittsfield, Massachusetts, built and tested several reproductions of the Khujut Rabu finds, all of which produced equivalent electrical output. Another German researcher, Dr. Arne Eggebrecht, found that his reproductions could electroplate selected items. Electroplating occurs when small electric current is applied to melt and adhere a thin layer of one metal, such as gold, onto the surface of another, such as silver. From his experiments, Eggebrecht believes that many classical statuettes and other objects regarded as solid gold are more likely gold-plated lead.

Existence of the Baghdad Battery and its companion pieces establishes that at least a fundamental understanding of electricity was grasped and applied by the ancients, even in a relative cultural backwater like Khujut Rabu during the third century BCE. Iran was then ruled by the Parthian Empire, a great military power, but not famous for its scientific sophistication. In any case, the batteries found there show that electrical power was not unknown in classical times, if not before. Rather than representing the beginning of a technology, the Baghdad Battery may have come near the end of developments with roots in the deep past, as indicated by a revealing comparison. The Pharos Lighthouse rose to 280 Old Kingdom Royal Cubits, or 481 feet, the same height as the Great Pyramid. Such a salient relationship was hardly coincidental, demonstrating that both structures, despite the millennia separating their construction, were built according to the same principles of sacred geometry.

This organizational unity began with all three pyramids of the Giza Plateau, which are linked by the Golden Section. Rediscovered by Leonardo da Vinci, who provided the name, the Golden Section is a spiral in the canon of ancient geometry used for the design of sacred architecture. It was valued as the most desirable proportion because it is expressed in the patterns of natural forms. These include cosmic nebulae, the ratios between planetary orbits, animal horns, sea mollusks, the formation of the human fetus, the laws of Mendelian heredity, heliotropism (the movement of flowers following the path of the sun), and whirlpools, together with thousands of other examples observed in nature. The Golden Section appears in a nautilus shell, its exterior wall removed to expose the spiral inside. This was the "Wind Jewel," the personal emblem carried by the Maya's Kukulcan and the later Aztec's Quetzalcoatl—the "Feathered Serpent," who long ago brought the principles of civilization to Mexico from his sunken kingdom across the Atlantic Ocean.

Temple was the first to notice that "a shadow cast by the second pyramid, known as the Pyramid of Khafre, upon the Great Pyramid at sunset on 21 December . . . if truncated by a vertical line running up the middle of the south face of the Great Pyramid, does actually form a golden triangle. There is actually a purposeful slight indentation of a few inches in the construction of the side of the pyramid, discovered in measurements made by Petrie. This 'apothegm,' as geometers call such vertical lines, forms the right angle to transform the solstice shadow into a perfect Golden Triangle." That this shadow was thrown on the Great Pyramid by Khafre each winter solstice to form " a perfect Golden Triangle" can have hardly been fortuitous, and further illustrates that all three pyramids were built simultaneously as part of a unified plan.

The evidence is overwhelming: the ancients possessed a technology in many instances equal to and occasionally more advanced than the vaunted efforts of modern humanity.

6 A Scientist Looks at the Great Pyramid

In a New Book, the Geologist Who Stunned the World by Redating the Sphinx Turns His Attention to Another Enigma Nearby

Robert M. Schoch, Ph.D.

Scrambling up a series of rickety wooden ladders tied together, I had to see for myself this "definitive proof" that the Great Pyramid was simply built as a tomb for Pharaoh Khufu (Cheops). Climbing nearly thirty feet above the floor of the Grand Gallery at its southern (upper) end, I crawled through an entrance barely two feet wide and on hands and knees made my way through a horizontal tunnel about twenty feet long and then up a series of smaller ladders, trying to keep the horrendous dust I could not help but kick up out of my eyes and lungs.

I was now climbing virtually straight up along a passage that had been blasted out with gunpowder in 1837 by the men working for the English explorer Colonel Howard Vyse. I passed a low-ceilinged chamber, with only enough room to squat, and then another one, and a third, a fourth, until finally I reached the fifth "secret" chamber, the one that Vyse had named "Campbell's Chamber." Here the roof is not flat and low like the other chambers, but you can actually stand up and stretch after slithering on your belly like a snake through the most narrow of openings blasted by Vyse in one corner.

Why had I come here? It was more than idle curiosity to see the chambers, which indeed many professional Egyptologists have never seen. I was on a quest to get to the heart of the meaning, the raison d'être, of the Great Pyramid. Were these simply "chambers of construction" or "relief chambers," meant to help support the enormous weight over the so-called King's Chamber? If so, why aren't there any relieving chambers over the Queen's Chamber, or the Grand Gallery, for that matter? Both of these structures, being lower in the pyramid, carry even more weight. Or are these somehow resonance chambers, a part of a giant machine that the pyramid forms? Are these the Hidden Heights, representative of the Halls of Amenti, the place of the Hidden God, as suggested by W. Marsham Adams back in 1895 and described in many of the ancient sacred texts?

Something that Howard Vyse had supposedly discovered in this uppermost chamber was a cartouche, the name of the king, which read "Khufu" roughly scrawled on the ceiling by some ancient hand in a red paint. The presence of the royal name of Khufu here was presumably the ironclad proof some Egyptologists had long sought to prove the Great Pyramid was nothing more than a huge mausoleum for the egomaniacal Fourth Dynasty pharaoh Khufu (Cheops), circa 2550 BCE. Making my way over the uneven blocks that compose the floor of Campbell's Chamber, using a copy of Vyse's original drawings as a guide, I found the long-sought cartouche in the back corner surrounded by hideous nineteenth- and twentieth-century graffiti. But the cartouche was there, sure enough, and it indeed read "Khufu"! So is this the end of the story? Are the traditional Egyptologists correct in their assertion that the Great Pyramid is nothing more than the gigantic tomb of Pharaoh Khufu? Maybe not. Indeed, on seeing the cartouche, I knew this was just the beginning of my adventure.

For one thing, this particular cartouche is turned up on end, and I would soon see in the other chambers that many of the red-painted inscriptions are completely upside down. What is going on here? Well, no one was meant to view these inscriptions once the pyramid was completed and access to

Fig. 6.1. Dr. Robert Schoch measures the alledged Khufu cartouche in the Great Pyramid. The cartouche was originally discovered by Howard Vyse.

these chambers cut off. Vyse had suggested they were nothing but "quarry marks" put on the blocks by the gangs that cut, hauled, and positioned the stone. But was Howard Vyse being totally honest? Had maybe his workmen who blasted and chiseled their way into these chambers in fact drawn these crude "Egyptian" inscriptions on the blocks themselves? Were these just fakes? Studying them closely, however, they looked authentically ancient to me. I could see later mineral crystals precipitated over them, a process that takes centuries or millennia, and the inscriptions continue under the overlying blocks. But there are more cartouches than the one of Khufu in the chambers.

Working my way down, sweating profusely and covered with grime, I explored the next chamber down—"Lady Arbuthnot's Chamber"—at length. Here are the most, if not best-preserved, cartouches—and not a single one says "Khufu"! Rather, here I found two different kinds of cartouches.

In one of the complete cartouches I could read "Khnum-Khuf," where "Khuf" or "Khufu" means "he protects me" and "Khnum" is the name of a god, so the whole name may be interpreted as "the god Khnum protects me." But who or what is being protected? Is it Pharaoh Khufu, or is the god Khnum actually protecting the Great Pyramid? Another complete cartouche has been interpreted simply as the name of the god, "Khnum."

Who or what was Khnum-Khuf? The early Egyptologist Sir Flinders Petrie suggested back in 1883 that maybe Khufu and Khnum-Khuf were co-regents who shared the throne of Egypt. Even more radically, it has been suggested that these cartouches are not even the names of a person or persons, but rather either different names for a single god or the names for several different gods. The researcher William Fix hypothesized,* based on the attributes of various gods, their symbolism, and etymological similarities, that "Khnum, Khnoum, Khufu, Souphis, Khnoubis, Chnouphis, Tehuti, Thoth, Mercury, Enoch, Hermes, and possibly 'Christos' are simply different representations of the same figure and power that finds remarkably similar expression in cosmologies extending over many thousands of years."

Is the Great Pyramid essentially the Book of Thoth memorialized in stone, as Marsham Adams contended? Did the postulant, the initiate, and the adept follow through the interior of the Great Pyramid to be subjected to trials of body and soul, ultimately (if successful) to die and be born again, finding illumination? Were the chambers of the Great Pyramid used for initiation rituals, just as the crypts and passages of the Temple of

---

*In his book *Pyramid Odyssey* (New York: Mayflower Books, 1978).

Hathor at Dendera or the Temple of Osiris would be thousands of years later?

I left the hidden heights of the Great Pyramid and was now determined to plummet its innermost depths. I was off to the Subterranean Chamber deep within the bowels of the rock under the Great Pyramid.

First, though, I found myself again in the inexplicable Grand Gallery. Nothing like it is found in any of the other pyramids of Egypt, or in the world, for that matter. Many explanations have been proposed for the Grand Gallery and the intricate internal geometry of the Great Pyramid. Was it an ancient power plant or a giant water pump? Was this all designed as a mechanism to interface or transition the pharaoh from life unto death to an eternal afterlife? Or was the Great Pyramid a product of the ancient Egyptian emphasis on the stars, where the Grand Gallery was at one time open at the top and could serve as a huge astronomical device used to observe the night sky in those pre-telescopic days? Or was it a gallery that displayed the images of the gods or the ancestors of pharaohs? Was it a hall of judgment where the postulant was questioned?

The mystery only deepens when we consider the narrow passages and two chambers, one off the top of the Grand Gallery and the other off the bottom, traditionally known as the "King's Chamber" and the "Queen's Chamber," respectively. Totally lined with granite shipped down the Nile from Aswan, the King's Chamber is not only a glorious site to the eyes, but a sensation to the ears as well. The acoustics and resonance qualities of the chamber, and indeed of the entire pyramid, are striking. To chant and meditate in the King's Chamber is a powerful emotional experience, one that I shared vicariously with Napoleon. It is said the dictator, ordering his aides away, asked to be left alone in the King's Chamber for the night. The next morning he emerged pale and shaken, refusing to his dying day to talk about what he had experienced there.

At one end of the King's Chamber sits a huge, lidless, solid granite "sarcophagus" or "coffer." This proves—many a traditional Egyptologist has claimed—that this is where the pharaoh was interred. Problem is, no riches or funeral trappings were ever found in the Great Pyramid. Tourists meditate and chant in the King's Chamber and, given the opportunity, will take turns lying in the granite coffer. It is an experience that is difficult to describe. I felt this was more a place of rebirth than death, and the granite box was more closely related to a baptismal font than a sarcophagus.

On the north and south walls of the King's Chamber are small openings that lead to narrow channels that run up and out to the exterior of the pyramid. Sometimes referred to as airshafts or ventilation shafts, it was long thought that they were simply functional, bringing fresh air to the chamber. But if it were a tomb, why would the deceased need all this fresh air? Was this room actually used by living people—perhaps for initiation or religious rituals? But what of the exactness and perfection of the shafts, pointing directly into the northern and southern sky? Four thousand five hundred years ago the northern shaft pointed to the star Thuban in the constellation Draco and the southern shaft pointed to the belt of Orion (associated with Osiris by the ancient Egyptians).

Next I made my way down the Grand Gallery and into the Queen's Chamber, which is smaller, and perhaps even stranger, than the King's Chamber. It has a gabled, pointed roof and a niche in one wall that mysteriously seems to reflect the cross section of the Grand Gallery. Was a statue or mummy placed here in the very heart of the Great Pyramid? Was it designed to hold a pendulum clock ticking away the moments of eternity? Or was it left empty, as the empty chamber in a Buddhist stupa, signifying the divine void that is all and nothing and is beyond mortal comprehension? And then there are the shafts that run from the north and south walls of the Queen's Chamber. The southern shaft apparently pointed toward the star Sirius (which the ancient Egyptians associated with the goddess Isis) in the ancient night sky while the northern shaft pointed toward the constellation we now know as the Little Dipper. In the case of the Queen's Chamber, there is no question that these shafts were ever used for the mundane purpose of ventilation. Not discovered until 1872, they originally stopped some inches behind the stone covering the walls. In the 1990s, robots were sent up them with small video cameras to explore, discovering small doors blocking the shafts! Later a robot drilled through one door, only to discover another door behind it.

Next I crawled down hundreds of feet through the Descending Passage to the chaotic Subterranean Chamber. Carved in the bedrock underlying the Great Pyramid, this room looks truly chaotic. Huge chunks of rock emerge from the floor, and there is also a strange "well" or "pit" on one end. Many traditional Egyptologists consider the Subterranean Chamber unfinished or abandoned. But why an unfinished chamber in what is arguably the most precisely aligned and built structure in the entire world? I received a powerful, strange sensation in this chamber. Other people have picked up powerful energies here also, as have inanimate machines!

At the Engineering Anomalies Research laboratory of Princeton University, since 1979 scientists have been running credible experiments on such much maligned subjects as extrasensory perception and the interactions of consciousness with matter. Sophisticated, highly sensitive, and finely calibrated electronic random event generators (REGs) can be used to measure—by the machines picking up anomalous nonrandom trends—the influence of mind over matter, that is, psychokinesis and the effects of consciousness on matter. Dr. Roger Nelson, an expert on the use of these REGs, carried a REG on a trip to Egypt in the 1990s and found many anomalies in the inner sanctums of various ancient temples. He also visited the Great Pyramid. Dr. Nelson found relatively little unusual activity in the King's or Queen's Chamber, but in the Subterranean Chamber the machine became very "excited."

Some researchers believe the Great Pyramid was built as a Freemasonic or Rosicrucian temple. Perhaps the "Rite of the Little Dead," during which the initiate spent three days in total darkness without food or water experiencing altered states of consciousness, took place in the Subterranean Chamber. Dr. Nelson's startling REG results are compatible with this hypothesis, as well as with a suggestion made by Robert Bauval, one I had independently been thinking about too. Perhaps the Subterranean Chamber, and the natural rock mound in which it is found—a rock mound that is now covered over by and enclosed in the Great Pyramid—is much older than the Great Pyramid itself. Was it considered sacred for thousands of years before the Great Pyramid was actually built?

It was getting late and the Egyptian authorities wanted me out of the Great Pyramid. Up I went through the Descending Passage and out to the dark, cool night sky with the lights of Cairo in the distance. My head filled with reverie, I began to think back to the late afternoon when I had first arrived at the foot of the Great Pyramid. Before entering I had examined the few remaining blocks of the once beautiful, highly polished, finely jointed façade. The four sides are aligned to the cardinal points with a degree of accuracy that is nearly impossible to duplicate today in a building of this size and mass. Many people do not realize it, but the four sides of the Great Pyramid are not perfectly flat; rather, they hollow slightly in, an effect that can be seen only under the right conditions. This hollowing may have been used in ancient days to determine the precise times of the equinoxes and solstices by observing the changing shadows on the sides of the Great Pyramid, an incredibly subtle and highly sophisticated methodology on a building whose base covers thirteen acres and that stands over four hundred and fifty feet high.

On the airplane from Cairo to New York the next day, my head filled with ancient dreams, I thought of what my Egyptian friend Emil Shaker had told me. Looking at a map of Egypt, the outlines of the country can be compared to a person, or more specifically the resurrected person in the form of Osiris, with raised and outstretched arms. The head of Osiris is the Great Pyramid, the body and legs are the Nile stretching to the south, the delta is the up-stretched arms that touch the Mediterranean Sea, which represents the sky. Osiris is both welcoming and gathering his children. The Great Pyramid beckons. Egypt draws one in. The quest had begun.

A not-so-final note: I had been to Egypt before, and since the trip I describe here I have traveled to Egypt on many more occasions. The quest is not yet over. After much meticulous study, I have concluded that the beginnings of the Great Pyramid extend back in time much earlier than generally thought, indicating a level of sophistication not usually acknowledged for such a remote period. In many ways, tracing the history and meaning of the Great Pyramid is key to understanding our origins as civilized beings. The Great Pyramid is not just a stagnant pile of ancient rock; it is a structure that embodies the human spirit, and it has lessons to teach us today.

7 Precession Paradox: Was Newton Wrong?

The Author of a Startling New Book Reexamines the Evidence

Walter Cruttenden

Stories about the "precession of the equinox" abound in the myth and folklore of ancient civilizations throughout the world. In the groundbreaking book *Hamlet's Mill*, authors Giorgio di Santillana, former professor of history and science at MIT, and Hertha von Dechend, of Wolfgang Goethe University in Frankfurt, detail the ancient fascination with the precession of the equinox. It is an exhaustive study that shows that prehistory peoples not only seemed to track the movement of the stars across the sky but related it to the rise and fall of the ages as well. Even Sir Isaac Newton wrote a little-known book, *The Chronology of Ancient Kingdoms,* wherein he attempted to match the world ages of the precession calendar with historical events. But why were our ancestors so obsessed with such an obscure astronomical motion? Today only a small number of astrophysicists try to understand the theoretical motions and dynamics of precession mechanics and they have completely disassociated its explanation with any myth or folklore.

Precession of the equinox defined the age-old phenomenon whereby the equinox moves backward through the constellations of the zodiac at the rate of approximately fifty arc seconds annually (one degree per seventy-two years). This means that an observer standing at the point of the equinox (the day when darkness and light are of equal length) who looks at the sky very closely will notice that exactly one year later (on the like equinox) the stars will not be in the exact same position as the year before. Because the exact timing of the point of equinox is so tricky and the movement is so small, it is quite difficult to detect in just one year, but over long periods of time it is unmistakable.

In 1543 Copernicus tried to explain this mystery and two others when he said that the earth had three motions. First, he said the sun appeared to move overhead from east to west not because the sun actually moved but because the earth spins on its axis. Second, he explained the seasons by showing that the earth went around the sun on a tilted axis (in a heliocentric system), thereby changing the length of the day and the amount of sunlight received. But he needed a third motion to explain the precession of the equinox. Here he postulated that the earth "librated," or wobbled. He assumed it was this wobble that changed the angle of the axis enough to cause the equinox to move or precess relative to the fixed stars. But he never said why it wobbled.

It was Newton, over one hundred years later, who—having just codified his laws of gravity—determined that the only bodies close enough or large enough to actually wobble the earth were the moon and the sun. Thus began the theory of "lunisolar precession" to explain the observable of the precession of the equinox.

PROBLEMS WITH LUNISOLAR THEORY

The "lunisolar" theory states that the earth's changing orientation to the fixed stars (primarily seen as the precession of the equinox) is principally due to the gravitational forces of the moon (luni) and the sun (solar) acting on the earth's equatorial bulge (the fat part around the middle). These objects are thought to produce enough opposing force or torque to slowly twist the earth's spin axis in a clockwise motion, so that after a period of approximately 25,770 years (at the current rate) the earth would have completed one retrograde motion on its own axis—and one retrograde orbit. In this theory, the earth is thought to behave like a wobbling top.

It is an observable fact that the earth's spin axis, and therefore the point of equinox, does change relative to the fixed stars. For thousands of years people have noticed the point of the equinox moving between the different constellations of the zodiac, and this is why it is said that we are now at the "dawning of the age of Aquarius." The vernal equinox is now leaving Pisces and moving into Aquarius, as seen on the first day of spring. The precession of the equinox is real when viewed against the backdrop of the fixed stars.

But here is the catch: There is no evidence that this observable is due to any change in the spin axis relative to the sun, or moon, or Venus, or anything "within" the solar system! Studies of the position of the axis relative to these bodies, just recently completed, confirm this. How can the earth appear to precess relative to objects outside the solar system but not relative to objects within the solar system? This is the precession paradox.

Remember that Copernicus guessed the spin axis must wobble but he never gave us a cause. It was Newton who assumed the earth wobbled relative to all objects, within and without the solar system, but he had no way to check if this was true.

As we now know, Newton's equations never did match observed precession rates, so along came Jean le Rond d'Alembert (1717–1783), followed by many others who have continually tweaked the formula to match observation. Ironically, none questioned the underlying assumptions (in science you usually don't question Newton). And so no one stood back to ask if this "wobble" might just be an apparent motion (one not occurring within our local reference frame of the solar system). To this day, astrophysicists continue to modify the calculations for precession, which now include many factors beyond the original "lunisolar forces" (including other planets, asteroids, a possible elliptical movement of the earth's soft core, and so on) all in an effort to better predict the precession rate. To me, this all looks suspiciously like a "plug," the act of coming up with new or different data to fit the predetermined answer. In the precession equation, the current answer is about 50.29 arc seconds per year of change in the earth's orientation to inertial space—it can be measured. Thus, a lot of inputs have been invented to get close to this answer. But all the "plugs" will never quite fit if the answer has a different cause.

The big thing wrong with the dynamicist approach (the process of looking strictly at the local gravitational dynamics) is the assumption that the earth's axis wobbles relative to all objects inside or outside the solar system. This is a blunder of historical proportions that has obfuscated our understanding not only of precession, but also of the very motions of the earth. Fortunately, new studies involving the timing of the Venus transits, lunar rotation equations, and the earth's motion relative to other objects in the solar system (such as the Perseids meteor shower) all show that the earth does not precess relative to local objects.

In spite of this, the current paradigm is so widespread and well accepted that when I mention the idea that the earth does not precess or wobble relative to local objects, astronomers are completely baffled, or they look at me as if I am insane. It is like telling people in Ptolemy's time that the sun does not go around the earth: they look up, see that it does, then conclude you're crazy. But the truth is, the so-called wobble is primarily the geometric effect of an unknown motion. There is an unaccounted-for reference frame—the solar system curving through space—producing the observable phenomenon we call precession.

LOOKING WITH NEW EYES

Here at the Binary Research Institute we have found that lunar rotation equations do not support lunisolar theory, nor does the earth's motion relative to nearby objects support the theory. Consider our largest meteor shower as a case in point.

As you may know, the Perseids are one of the most reliable meteor showers of the year. Caused by the Swift-Tuttle comet debris slicing through the earth's orbit path, it peaks on August 11–12 each year (depending on recent leap year adjustments). Hence, it effectively acts like a marker that intersects the earth's path around the sun, and has been plotted for eons. The very best records go back to at least the Gregorian calendar reform of 1582—a time from which we know we have a highly reliable calendar system (less than one day of error per thirty-two hundred years).

But here is the issue: According to lunisolar theory, the earth does not go 360 degrees around the sun in a tropical or equinoctial year—it has to come up fifty arc seconds short because that is the amount of precession we can measure relative to the distant stars. Because the tropical year is so close to the average calendar year, objects in space appear to slip through the calendar at the rate of about one day per seventy-two years. Now if precession is caused by local forces, we would expect the observation date of the Perseids to change at the same rate that the earth precesses relative to the fixed stars outside the solar system. This means that just as the constellations have changed position relative to the equinox at the rate of about one degree or one day every seventy-two years, or almost a full week since the Gregorian calendar reform, the Perseids (debris within the solar system) should have done the same (relative to a wobbling earth). This means that the shower should now be peaking around August 5 or earlier. But the fact is the Perseids have moved very little if at all in those 423 years. This meteor shower is even called "St. Lawrence's Tears" because it happens regularly right after his feast day. Why hasn't it precessed through the calendar like everything else outside the solar system?

It is possible that this comet debris just happens to drift in an opposite direction about the same rate as precession and that our lunar calculations and Venus transit calculations are also somehow precisely incorrect, but I don't think so. A more logical conclusion is that we can't measure precession relative to objects within the solar system because it (the precession observable—the earth's changing orientation to inertial space) is not caused principally by local forces. Yes, there are some local forces and they produce the minor motions of nutation, Chandler movement, and the like, but the major change in orientation that we experience (at least relative to the fixed stars) is likely not due to any large local wobbling of the axis— but rather to the entire solar system (a moving reference frame in itself) gently curving through space. It produces the observable without the need for significant local force. That is the only way I know of solving the paradox of an earth that does not change orientation relative to local objects within the solar system while clearly changing orientation to objects outside the solar system in excess of fifty arc seconds per year.

In addition to the Perseids data, we have found that precession is actually accelerating and acts more like a body that follows Kepler's laws (in an elliptical orbit) than a wobbling top that should be slowing down. Furthermore, there are at least half a dozen circumstantial arguments indicating that precession is a result of something other than local forces.

We are not the only ones to make these observations. A number of completely unconnected groups, including Karl Heinz and Uwe Homann at the Sirius Research Group in Canada, have come to the same conclusion: The lunisolar theory of precession does not make sense. Their Venus transit studies show that either the nodes of Venus just happen to match the precession rate (a highly unlikely scenario) or the earth does not wobble relative to Venus. And studies of the motions of the moon relative to the earth show the same thing—there is no precession of the earth relative to the moon.

If this was any easy problem to understand, I am sure it would have been corrected by now. But it is an extremely tricky thing to try and measure any change in the earth's orientation relative to objects within the solar system because everything nearby has such a high relative motion—everything is moving! This is why astronomers use very distant objects, quasars in other galaxies, when measuring changes in the earth's orientation (precession). But such measurements will never show how much the earth's spin axis changes relative to local objects—it has just been assumed the change is the same. That is the problem: incorrect assumptions.

If the observable of precession is the result of the solar system curving through space rather than a local wobbling of the axis, the big question is: What causes the sun or solar system to curve through space?

THE BINARY HYPOTHESIS

If our sun is part of a binary (or multiple star) system, it would be gravitationally bound to a companion star, resulting in the sun's curved motion through space around a common center of gravity. This is the accepted motion pattern of binary star systems: two stars attracted to each other orbiting a common center of mass or gravity.

This motion, combined with an oblate earth that is subject to even minor local torque (gravitational effects like lunisolar forces on a very small scale), would cause a constant reorientation of the earth's spin axis relative to inertial space, commensurate with the motion of the binary. Thus, if the binary motion caused the sun to circle the center of mass in 24,000 years, then the spin axis would appear to reorient itself to inertial space in this same period (plus or minus any purely local effects). This principle works because the local motion occurs within the confines of the binary movement allowing the binary movement to distort whatever local motions are actually occurring. In this case, the observable of precession would be due principally to the geometric effect of a solar system that itself curves through space (around the binary center of gravity). The solar system here acts as a distinct reference frame that contains all the motions of the planets and their moons, which in turn maintain all their respective gravitational relationships, as the system as a unit moves in a spiral motion relative to inertial space, similar to the way a galaxy appears to move as a unit relative to inertial space.

In simple terms this means that the earth doesn't really wobble very much, at least within the reference frame of the solar system. It just looks like it is wobbling relative to the fixed stars because the whole solar system is moving—another reference frame is at work.

BINARIES EVERYWHERE

It is important to note that there was little or no knowledge of the extent of binary star systems at the time the current lunisolar model was put forth in the West. Even when I was a boy in the 1950s and '60s dual star systems were thought to be the exception rather than the rule. However, it is now estimated that more than 80 percent of all stars may be part of a binary or multiple star relationship. Apparently, stars like companions as much as people do! Since we now know that numerous star types such as black holes and neutron stars and many brown dwarfs (and even red dwarfs against the galactic center) are almost impossible to see and very often difficult to detect, the number of multiple star systems may be higher than a census of strictly visible stars would indicate. Thus, if most stars out there have companions, our lone sun and its solar system are looking more and more like an anomaly—that is, if indeed it is a single star system, and not a partner in a multiple star system.

Assuming that we are in a binary system, and that Newton's laws work just as well outside the solar system as inside it, then the sun's dual would most likely need to be a dark companion such as a brown dwarf, or theoretical old neutron star, or even some large planetlike mass that also has a very long orbit period (making any of its effects difficult to notice). It could even be a not-too-distant black hole that is not currently consuming matter and therefore is difficult to detect, though this is highly doubtful.

Another possibility is that MOND (Modified Newtonian Dynamics) or some variation of local gravitational dynamics comes into play at long distances outside the solar system. This of course would open the possibility that the sun has a visible companion (and coincidentally would solve much of the dark matter problem). We can't expound on this particular possibility without significant further research, but we can't rule it out either, given the growing evidence that something is moving our solar system in an elliptical pattern far tighter than any galactic motion would produce. My gut feeling is that we have a lot to learn about the subject of gravity and gravitational tides. Right now there is a lot of extremely interesting new research going on that could greatly expand our companion star possibilities.

SO WAS NEWTON WRONG?

Copernicus and Newton were brilliant scientists and far ahead of their time. Given the fact that they were doing their work in a period when the acceptance of a heliocentric system was still in question, it is understandable that they could not figure out the third motion of the earth. For Copernicus to say the sun does not move in his first motion and then say it does to produce the third motion would have been too much to ask. Likewise, for Newton to deduce that the whole solar system was curving through space, meaning the sun was moving, before it had been accepted that any star could move, would also have been an overly ambitious thought. Besides, no one knew of the prevalence of binary stars or stellar dynamics of any type in that early period. So Newton is off the hook.

But for our modern astrophysicists to continue to assume that the earth does not change orientation relative to objects within the solar system any differently from how it does relative to objects outside the solar system is unacceptable. We now have the tools to differentiate and it is time to study more thoroughly the earth's motion relative to all objects.

The ancients hinted at a lost star in their myth and folklore, and they implied it drove the rise and fall of the ages. If we discover we are in a binary system, with waxing and waning influences from another star, who knows, we might just prove the ancients right!

8 The Dogon as Physicists

Do the Symbols of This Enigmatic African People Show Knowledge of Theoretical Physics?

Laird Scranton

In the decades since the appearance of Robert K. G. Temple's book *The Sirius Mystery*, there has been ongoing debate about the cosmology of the Dogon tribe of Mali. The key question has centered on the primitive tribe's seemingly anomalous knowledge of details relating to the star system of Sirius—knowledge that some use to support the claim of an alien contact and that others see as, at best, information implanted with the tribe by some enlightened outsider. Attempts to refute the alien contact claim have gone so far as to call into question the methods used during the 1940s by French anthropologists Marcel Griaule and Germaine Dieterlen in their decades-long study of the tribe—the study that sparked Temple's initial interest in the subject.

There are other attributes of the Dogon tribe that contribute to the enduring popularity of the Sirius question. For one thing, Dogon mythology includes a host of symbols and stories that bear strong resemblance to the ancient Egyptian religion, as has been noted by scholars such as Nicholas Grimal. At the same time, Dogon religious rituals share many of the trappings of early Judaism, such as the practice of circumcision and the traditional celebration of a jubilee year every fifty years. These similarities cause some to ask whether Dogon knowledge should be interpreted as a remnant of a very ancient tradition of knowledge or merely as an intrusion of more modern knowledge.

At this point in time, the answers to Temple's mysteries can no longer be reasonably found in the Sirius question itself, because the debate has succeeded in casting doubt on so many of Temple's assertions. However, there are many other fascinating aspects of the Dogon religion and cosmology not entangled in this debate that seem to have been overlooked for study, due to the continuing glare of the Sirius star. Some of the most promising of these are the Dogon symbols relating to the structure of matter.

Dogon mythology, like many of the most ancient mythologies, describes the formation of the universe from an egg containing all of the seeds or signs of matter. This description is not unlike the typical scientific description of the unformed universe prior to the Big Bang. The Dogon say that a spiraling force within this egg caused it to open, releasing a whirlwind, which ultimately gave rise to the spiraling galaxies of stars and planets. The wind itself was said to be the Dogon's one true god, Amma. The first finished creation of Amma was a tiny seed called the *po*. The Dogon describe this seed in terms quite reminiscent of the atom—they say that Amma creates all things from the cumulative addition of like elements, beginning with the po.

According to the Dogon, the po itself comprises components called sene seeds. The sene are described by the Dogon in terms that bring to mind protons, electrons, and neutrons. The sene combine together at the center of the po, much like protons and neutrons in the nucleus of an atom, and then surround it and give it form and visibility by crossing in all directions, much like electrons orbiting a nucleus. The Dogon represent the sene with a drawing that looks like four oval petals of a flower arranged together to form the shape of an X. An intriguing aspect of this drawing is that it closely matches one of the most common shapes that electrons form when they orbit an atom.

The Dogon discuss the formation of the sene seeds themselves—the germination of the sene—a process that the Dogon represent with yet another drawing. This drawing consists essentially of four circles with varying numbers of "spines" sticking out. One circle has four spines, one has three spines, another two—all symmetrically arranged. The last circle contains a haphazard assemblage of spines in no particular arrangement. To make sense of this drawing, it is necessary to understand a little about quantum particles—the building blocks of electrons, protons, and neutrons—and how they are categorized by modern science. Each quantum particle has a property called "spin," which essentially tells us what the particle looks like from different directions. Scientists group quantum particles into four categories based on this "spin" property. Particles in the first category look the same from all angles, like a globe. Particles

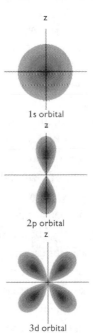

z

1s orbital

z

2p orbital

z

3d orbital

Fig. 8.1. Different views of the sene seeds that make up the po; its structure is very similar to that of an electron.

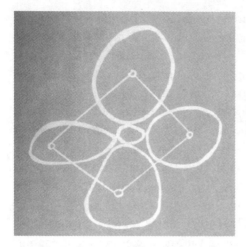

Fig. 8.2. This drawing illustrates the germination of the sene, which closely resembles the four quantum spin categories.

in the second category are like an arrow—they must be turned 360 degrees in order to look the same. Particles in the third group are like a double-headed arrow. They must be turned 180 degrees to look the same. Particles in the last group are the most difficult to conceptualize. Contrary to logic, they must be turned around twice in order to look the same. Remarkably, the figures in the Dogon drawing of the germination of the sene closely resemble and describe these four quantum spin categories.

Since much of modern quantum science is still theoretical and is only in the process of being experimentally proved, no one can definitively state the exact number of quantum particles that might exist. However, a good estimate is that there are more than two hundred fundamental particles. Dogon mythology, on the other hand, reflects a more precise understanding—it defines 266 elementary seeds or signs.

In order to understand the component structure of quantum particles, we must turn to the science of string theory. String theory came to the forefront of scientific thought in the early 1980s. It proposes that the smallest components of matter consist of tiny one-dimensional loops that vibrate like rubber bands at varying rates. These vibrations, in turn, give rise to the different types of quantum forces and particles. At this time, string theory remains unproved, primarily because the component strings are many times smaller than the smallest particle scientific technology can image.

According to string theory, one function of quantum strings

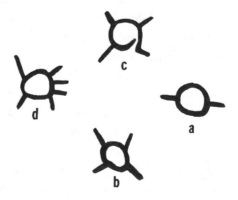

Fig. 8.3. Could these Dogon symbols represent the four fundamental forces?

is to give rise to the four quantum forces—the gravitational force, the electromagnetic force, the strong nuclear force, and the weak nuclear force. Under some circumstances, these one-dimensional loops are also thought to be able to join together to form two-dimensional membranes.

Annually the Dogon perform the ritual drawing of a figure on the ground to represent the 266 seeds or signs of Amma. This drawing consists of a small circle within a larger circle. The space between the circles is filled with a series of zigzag lines. When the figure is complete, the Dogon say that the signs have been drawn. The completed figure bears a strong resemblance to the scientific diagram of one of the typical vibrational patterns of a quantum string.

For the Dogon, these 266 elemental signs are the work of the spider of the sene whose thread—much like the quantum string—is said to "weave the words" of the signs. Unlike the theoretical quantum loop, the Dogon say that this thread is coiled, much like a spiraling galaxy. The thread of the Dogon also has the capability of forming a thin skin or membrane—one compared by the Dogon to the thin covering on the outside of the brain. The thread also gives birth to four seeds, similar to the four quantum forces, whose names in the Dogon language mean "to draw together" (gravitational force), "bumpy" (electromagnetic force), "stocky" (strong nuclear force), and "that bows its head" (weak nuclear force).

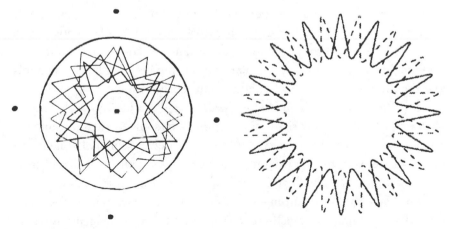

Fig. 8.4. Ritual Dogon drawing of a figure on the ground, which represents the 266 seeds or signs of Amma.

Fig. 8.5. One of the typical vibrational patterns of a quantum string. (Reprinted from The Elegant Universe by Brian Greene, copyright © 1999 by Brian R. Greene. Used by permission of the publisher, W. W. Norton & Company, Inc.)

In essence, Dogon religious mythology appears to accurately describe the true underlying structure of matter, organizes it in the proper sequence, diagrams it correctly, and ascribes to it the correct attributes of each component. Since the symbols belong to the mythology of an outwardly primitive African tribe, there would be no reason for an anthropologist to compare what seem to be simple tribal drawings to esoteric scientific diagrams. But when the comparison is made, we see that the result is a very close match.

Discussion of the Dogon symbols relating to the structure of matter turns out to be a much simpler prospect than that of the stars of Sirius. For one thing, we can test the symbols against a fixed standard—either they match the scientific structure of matter or they don't. Moreover, there is no question of implanted knowledge, because much of the deep science needed to understand the Dogon descriptions did not even enter the modern consciousness until after 1980, so is not likely to have been implanted with the tribe decades earlier by Griaule or Dieterlen.

A careful study of the Dogon creation story and the anthropological studies of Marcel Griaule and Germaine Dieterlen offers a wealth of intriguing insight into the possible underlying wisdom of the Dogon, presented in terms that often seem to make complete scientific sense. As one becomes more familiar with it, the Dogon creation story reveals itself as a carefully organized presentation of ideas and symbols relating to the creation of the universe, the creation of life, and the creation of civilization. One need not postulate an alien contact to perceive it, nor can any speculation about implanted knowledge explain it. Furthermore, just as it has been possible to show a correlation between Dogon symbols and the components of matter, it is equally possible to show a similar relationship between Dogon symbols and the components of genetics and human reproduction.

Most importantly, these Dogon symbols and stories may well contain gigantic hints about the origins and meanings of some of the most ancient religious symbols and stories, which they so closely resemble. Parallels can be drawn between many of the most important Dogon concepts and their clear counterparts in the Egyptian religion in its earliest form. For example, the word *po*—the Dogon atom—sounds very like the Egyptian hieroglyphic word Pau (the name of a self-created god) and a related Egyptian word, *pau-t,* which means "matter or substance." All of this supports the conclusion that further study of the Dogon culture might well provide an important template for our understanding of a wide variety of modern anthropological, archaeological, scientific, and religious issues.

9

The Astronomers of Nabta Playa

New Discoveries Reveal Astonishing
Prehistoric Knowledge

Mark H. Gaffney

According to most experts the dawn of Western civilization occurred in the fourth millennium BCE with the sudden flowering of Sumer in southern Iraq and pharaonic Egypt soon after. This is the mainstream view that was taught when I was a college student. Increasingly, however, it is under assault. Recent discoveries are challenging almost everything we thought we knew about human history. In 1973 a team of archaeologists made such a discovery while traveling through a remote region in southern Egypt. They were navigating by compass through a trackless waste known as the Nabta Playa and had halted for a water break when they noticed potsherds at their feet. Fragments of old pottery frequently are an indicator of archaeological potential, and the team returned later to investigate. After several seasons of digging they eventually realized that Nabta Playa was not just another Neolithic site. The breakthrough came when they discovered that what had looked like rock outcroppings were in fact standing megalithic stones.

They also found a circle of smaller stones, which in photos look like derelict rocks. Nearby, the arrangement of larger megaliths sprawls over a broad area. The windswept site is desolate beyond belief. But thousands of years ago this forbidding waste was a well-watered grassland and seasonally, at least, well peopled.

Today we know that the great megaliths of Nabta Playa are anything but random stones. Long ago, someone relocated them from a still unknown quarry—but for what purpose? Subsequent excavations led by Fred Wendorf, one of the discoverers and a much traveled archaeologist, turned up an abundance of cultural artifacts, which were radiocarbon dated. The ages ranged from 10,000 BCE to 3000 BCE, with most of the dates clustered around 6000 BCE, when the climate was much wetter than now. The Nabta Playa is a basin, and during this epoch it filled with seasonal lakes. Excavations through the eight to twelve feet of sediment laid down during this period showed that some of the megaliths had been buried intentionally. The team also found strange carvings in the bedrock under the sediment—evidence of great antiquity.

The archaeologists mapped the area and used global positioning technology (GPS) to plot the locations of twenty-five individual megaliths. Many others remain to be plotted. Fortunately, the site's remoteness protected it from most human disturbance. Though the mapping data hinted at astronomical significance, Wendorf's team searched in vain for the key to unlock the site. In 2001 they presented their research in a book edited by Wendorf, *Holocene Settlement of the Egyptian Sahara*. The two-volume study makes for interesting reading. But its authors had few answers.

However, even as Wendorf's book was in press, a former NASA physicist named Thomas Brophy was quietly pursuing his own astronomical study of Nabta Playa. Brophy had already reviewed the sparse data published in *Nature* in 1998, and after Wendorf's more extensive data became available his nascent theories fell into place. In 2002 Brophy presented his findings in *The Origin Map*. Because the available astronomy software was inadequate, Brophy had to custom-engineer his own. Thus armed, he was able to track

Fig. 9.1. A very finely worked Calendar Circle stone (about 2 feet long) that has toppled out of its place. It appears to be extremely hard, like flint, and shows very little indication of weathering. (Photograph courtesy of Thomas Brophy)

Fig. 9.2. The other end of the Calendar Circle stone shown in the
photograph on page 64. (Photograph courtesy of Thomas Brophy)

star movements at Nabta Playa over thousands of years, and succeeded in
decoding the stone circle and nearby megaliths. The "Calendar Circle" has
a built-in meridian line and a sight line—both conspicuous—which indicated to Brophy that the circle was a user-friendly star-viewing platform. Its
design was so simple that even a novice could have used it. A night viewer
between 6400–4900 BCE stood at the north end of the meridian axis and
allowed himself to be guided by three stones at his feet to the constellation
Orion overhead. The correspondence between ground and sky would have
been self-evident: The three stones within the outer circle are laid out in the
precise pattern of the stars of Orion's famous belt, before summer solstice,
as indicated by the Calendar Circle itself. Once the pattern becomes familiar, it is unmistakable.

In another section of his book Brophy concluded that Robert Bauval
and Adrian Gilbert were at least partly correct in their 1992 study, *The
Orion Mystery,* in which they claimed that Giza had a similar planned structure. Bauval and Gilbert argued that the pyramids of Giza were constructed
to mirror heaven, laid out on the ground to represent these same three stars
of Orion's belt.

Here, then, at Nabta Playa, was evidence of a common astronomical tradition of astonishing longevity. Just to give you some idea, modern astronomy
is about five hundred years old; the astronomy common to both Giza and

Nabta Playa survived for at least six to seven thousand years, possibly longer. The shared astronomy also suggests a shared cultural tradition. In fact, Wendorf's team amassed considerable evidence of overlap between the Neolithic Nabta culture and the much later Old Kingdom of pharaonic Egypt, when pyramid building reached its zenith. It is interesting that more than a century ago Flinders Petrie, one of the founders of Egyptology, arrived at a similar conclusion. He found evidence that the enigmatic Sphinx was not an Egyptian sculptural form at all but rather had originated in Ethiopia.

Brophy's findings also support the work of geologist Robert Schoch, who recently found telltale evidence of water erosion on the Sphinx indicating that the most enigmatic sculpture on the planet dates to this same wetter epoch or before. Schoch's analysis flew squarely in the face of mainstream Egyptology, which continues to insist on a much later date. A confirmed link between Giza and Nabta Playa would lay to rest any doubts about the relevance of Nabta Playa because of its remoteness. Far from being disconnected from the mainstream of the emerging Egyptian cultural tradition, at one time Nabta Playa may even have been the center of it all.

While all of this is extraordinary, Brophy's conclusions about the other nearby megalithic formation are mind numbing. Brophy thinks this other construction may be a star map, the creation of which required a knowledge of astronomy that rivaled and may even have surpassed our own. Brophy's conclusions are highly controversial, but his work deserves close attention because if he is correct, we have barely begun to understand where we came from.

So, what do the Nabta megaliths tell us after thousands of years of silence? Their designers placed them in straight lines that radiate from a central point. The arrangement employed a simple star-coordinate system that assigned two stones per star. One aligned with the star itself and marked its

Fig. 9.3. *The physicist Thomas Brophy beside megalith "A-O." (Photograph courtesy of Thomas Brophy)*

vernal equinox heliacal (that is, rising together with the sun on the first day of spring) position on the horizon. The other aligned with a reference star, in this case Vega, thus fixing the first star's rising at a specific date in history. In archaeoastronomy, single megalithic alignments with stars are considered dubious because at any given time several stars will rise at or within a few degrees of the point on the horizon denoted by a lone marker. Over long periods of time, many different stars will rise over this position. The creators of Nabta Playa eliminated uncertainty with the Vega alignment and the specificity of vernal equinox heliacal rising, which occurs only once every 26,000 years for a given star. This fixed the star's rising date. Vega was a logical choice because it is the fifth brightest star in the heavens and dominated the northern sky in this early period. Brophy found that six of the megaliths corresponded with the six important stars in Orion (Alnitak, Alnilam, Mintaka, Betelgeuse, Bellatrix, and Meissa), also confirming his analysis of the nearby circle. Their placement marked the vernal heliacal rising of these stars, which occurred around 6300 BCE, within about twenty years. The second set of reference stones were keyed to the heliacal rising of Vega, which occurred at the autumnal equinox. In the seventh millennium BCE the Nabta plain was a busy place.

The heliacal rising of a star occurs when it rises above the horizon with the morning sun. A vernal heliacal rising describes the same event on the day of the spring equinox, which is much rarer. Using a conservative statistical protocol, Brophy calculated the probability that the megalithic alignments at Nabta were random at less than two chances in a million, which, as he writes, "is more than a thousand times as certain as the usual three standard deviations requirement for accepting a scientific hypothesis as valid." The only reasonable conclusion is that the star alignments at Nabta Playa were carefully planned—no accident.

But hold on, because from here Brophy ventures into wild territory. He was puzzled by the fact that the Nabta megaliths were not placed at uniform distances from the central point. Brophy writes: "If the varied distances didn't have a purpose, one would expect the skilled Nabta Playa designers to have used a more pleasing arrangement . . . [therefore] the distance placements are suggestive of a meaningful pattern." Students of the Giza Plateau have often remarked that no detail of the famous pyramids was left to chance. Every angle, every relationship, every aspect, had a definite purpose. Brophy merely guessed that the same might hold at Nabta Playa. So, what did the variable distances of the megaliths from the central point represent? After considering a number of alternatives, just for fun Brophy theorized: What if the distances on the ground were proportional to the actual distances of the stars from earth? When he looked up the current best measurements as determined by the Hipparcos Space Astronomy Satellite, Brophy was blown away. They matched in each case to within about a standard deviation. The proportional scale turned out to be one meter on the ground at Nabta = .799 light-year. The match is "more than astonishing," as Brophy writes, because even with modern technology the science of measuring the distances to stars is a tricky and imperfect business. Current best measures of distance must be regarded as approximations. Brophy's conclusion bears repeating: "If these star distances are the intended meaning of the Nabta Playa map, and are not coincidence, then much of what we think we know about prehistoric human civilizations must be revisited."

Brophy believes information about the relative velocities of stars, and their masses, may also be encoded in the placements. And he thinks that smaller companion stones lying near the base of some of the large megaliths probably represent companion stars, or even planetary systems. Unfortunately, this cannot be tested at present because astronomy is not yet able to observe earth-sized planets across the reaches of space. Rapid strides are being made, however. A number of Jupiter-sized giants have already been detected and resolving power continues to improve. Soon we may know if Brophy's staggering idea is correct.

A GALAXY MAP?

Nabta Playa held other surprises. The location of the star map's central point initially drew the attention of Wendorf's team because a complex structure of megaliths had been placed there. One large stone stood squarely at the central point, surrounded by others. Numerous other stone complexes had also been placed in the vicinity. These appeared to be burial mounds and when the archaeologists excavated two of them, the team expected to uncover mortuary remains. Instead, they dug through twelve feet of Holocene sediments to bedrock and found bizarre carved sculptures, which they never did explain.

Later, Brophy examined these in light of the deciphered star map and was blown away again. He realized that whoever created Nabta Playa might have been in possession of advanced knowledge about our Milky Way galaxy. The bedrock sculpture appears to be a made-to-scale map of the Milky Way as viewed from the outside—that is, from the perspective of the north galactic pole. The map correctly indicates the position, scale, and orientation of our sun, and the placements of the spiral arms, the galactic center, even the associated Sagittarius dwarf galaxy, which was discovered only in 1994. Although Wendorf's excavation had dismantled the stone complex on the surface in the process of exhuming the underlying sculpture, Brophy was able to determine from Wendorf's accurate diagrams/maps that the central point was directly above—and surely represented—the correct position of our sun on the galaxy map.

Brophy then made another key discovery: One of the megalithic sight lines stood in relation to the galactic center. Its alignment marked the galactic center's vernal heliacal rising circa 17,700 BCE. Amazingly, the orientation of the galactic plane in the sculpture also jibed with this date. Brophy concluded that the stone sculpture was a map of the Milky Way as seen from the standpoint of the northern galactic pole. Next, he turned his attention to the second stone complex excavated by Wendorf's team, which, likewise, had produced no mortuary remains. Its size and placement suggested that it was a map of Andromeda, our sister galaxy. Calculations showed that its size—about double the Milky Way—and its placement may be consistent with Andromeda's known size/location.

As for Nabta Playa's other stone complexes, they have not yet been investigated . . .

GIZA: A PRECESSIONAL CALENDAR?

Brophy also conducted his own independent investigation of Giza, and there too found evidence the designers knew about the galactic center. Brophy's powerful software enabled him to refine Robert Bauval's estimate of the Great Pyramid's correlation date. Brophy agrees that the famous star shafts serve as markers that fix the date of correlation to within a narrow window. When Brophy precessed the sky above Giza he found that the best shaft alignment occurred around 2360 BCE—about a half-century later than Bauval's date. Bauval argued that the southern shaft of the King's Chamber aligned with Orion at the time of construction. But Brophy found that the last of the three stars of Orion's belt, Alnitak (Zeta Orionis) aligned with the southern shaft more than a century before. The southern star shaft's alignment with the galactic center at the time of pyramid construction also corroborated his findings at Nabta Playa.

Assuming that the layout at Giza is a mirror of Orion's belt, when exactly did this occur? Bauval's preferred date is 10,500 BCE, long before the pyramids were actually built, when the three stars of Orion's belt reached their southern culmination of the 26,000-year precessional cycle. When Brophy tested this idea, however, he discovered another layer of complexity. He found that the ground mirrored heaven on two dates: in 11,772 BCE and again in 9420 BCE. Brophy concluded that the construction was never intended to designate Orion's southern culmination, as Bauval argued, but rather to bracket the epoch in which this occurred. The two dates also bracket another important event, the northern culmination of the galactic center at around 11,000 BCE. In other words, Giza was constructed as a zodiacal clock, set in stone to the grand precessional cycle. This supports the view that the site's astronomy long predated the actual pyramid construction.

Fully cognizant of the revolutionary nature of his analysis, Brophy wisely makes no final pronouncements in his book. He merely presents his findings as a working hypothesis and invites others to investigate further. Fortunately, many of his ideas are testable. So far, only twenty-five Nabta stones have been plotted with GPS, and only two of at least thirty stone complexes have been excavated. Time will surely tell. . . .

PART THREE

CHALLENGES TO
TRADITIONAL PHYSICS

10 Tesla, a Man for Three Centuries

Our Debt to the Eccentric Croatian Inventor Continues to Grow

Eugene Mallove, Ph.D.

Near midnight between the ninth and tenth of July 1856, Nikola Tesla was born of Serbian parents in Croatia near Bosnia, an area that has known centuries of turmoil. From such a humble beginning came a "Man Out of Time"—almost literally—the title of Margaret Cheney's 1981 biography of Tesla. The eccentric genius who would grow from this infant was a man rooted in the nineteenth century's scientific revolutions of electricity and magnetism. His electrical creations would transform the twentieth century beyond recognition—widely distributed electricity that would dominate all aspects of life and a society pervaded by global "etheric wave" communication, radio, and television. Nikola Tesla's legacy has not yet come to full bloom, but it surely will in this, the twenty-first century, which (one hopes!) will be much less recognizable to twentieth centurions than the twentieth century was to the Victorians.

Tesla came to the United States in 1884 with a letter of introduction to Thomas Edison—from Charles Batchelor, the British engineer who ran the Continental Edison Company in Europe. The two towering figures, Edison and Tesla, had a very brief working relationship in America, but their manners, personalities, and approaches to commercializing the generation and transmission of the electrical "fluid" clashed dramatically. Edison would remain stuck with the problematic direct current paradigm, while Tesla had long envisioned his polyphase alternating current approach, which he invented down to its many particulars. Tesla won, of course, yet in 1943 he died in the New York hotel room in which he lived, in debt and in abject poverty. He was a visionary scientific genius, not a savvy, cutthroat businessman. To this day, Tesla does not receive the credit he deserves. Though he was the true discoverer of the basic methods of radio communication (as later formally adjudicated in a U.S. Supreme Court decision following Tesla's death), popular history grants the honor of radio's invention to Marconi, who had used Tesla's ideas. Tesla knew of that interference at the time, but he just smiled; he was too immersed in his other overarching plans for energy and communication.

We are far from the time of nineteenth-century and early-twentieth-century Nikola Tesla, when experiment and the concrete technological device based on experiment was the ultimate arbiter of Truth. Today we live in establishment science's world of Absolute Fiction. In this world, hundreds of experiments that show irrefutable evidence of nuclear reactions occurring at low-energy (LENR/"cold fusion") can be dumped into the trash bin of alleged scientific failure—to cite one example of many such trashings. When Tesla worked, another very serious component of the universe was very much a critical topic of discussion among physicists—the aether (or ether). This was the postulated finely structured substance that simply had to exist if there was to be any hope of explaining how light waves were to travel through what otherwise would be just a vacuum of absolute "nothingness."

It is the opinion of conventional science that the universe consists of just this: mass (embodied in particles such as electrons, protons, neutrons, and various antimatter particles) and electromagnetic radiation (visible light, radio waves, ultraviolet and infrared radiation, X-rays, gamma rays, and so on). All of this "stuff" is embedded, as it were, in what is called a space-time plenum, which supposedly emerged into existence—from previous nonexistence—in a fraction of a second of "cosmic time" some 15 billion years ago (declared "Confirmed"! by the *New York Times* on February 12, 2003, as 13.7 billion years ±200 million years). For some reason, universal "cosmic time" is all right to talk about, as distinct from curved or flat "space-time," which we mortals must be confined to, according to Einstein's Special Theory of Relativity; we cannot have our time and our space separately, or so we are told: no one's time is the same time as that of another, who is moving relative to him or her.

In conventional science, all the "stuff" of the universe fills a regime of cosmic nothingness, with quantum mechanical electromagnetic fluctuations at an extremely small subatomic level filling up this "nothingness"—the so-called zero-point energy. Virtual particles supposedly pop in and pop out of existence—unpredictably, chaotically, randomly—to satisfy or not satisfy mass-energy conservation. Recently more baggage has been added to this cosmic picture by conventional science. It feels a need to augment the universe with so far unidentified "dark matter," "dark energy," "quintessence," and a seemingly interminable epicyclic bestiary of imagined creatures to help patch up the Big Bang with its primary structural feature, curved space-time, as dictated by general relativity. This is Einstein's theory that supposedly "explains" gravity, but which does no such thing.

An all-pervading "dark energy" is the latest Establishment darling that is supposedly accelerating the imagined expansion of the universe. Said expansion is critically dependent on the measured cosmic redshifts in light being truly of cosmic significance, not merely local-to-the-observed galaxy or quasar (meaning they are of new physics significance)—but that is another depressing story of the failures of "Fizzix" to consider possible major cracks in its foundations.

James Clerk Maxwell, who gave us the first version of the equations used in electromagnetic theory today, certainly believed in an aether—the luminiferous, static one that carried light. As he wrote in the ninth edition of the *Encyclopaedia Britannica* (which began appearing around 1875), "The only aether which has survived is that which was invented by Huygens to explain the propagation of light. The evidence for the existence of the luminiferous aether has accumulated as additional phenomena of light and other radiations have been discovered; and the properties of this medium, as deduced from the phenomena of light, have been found to be precisely those required to explain electromagnetic phenomena. "By the time of the eleventh edition of *Britannica* (1910), the aether was still alive and well—five pages of fine print and mathematics in the first volume discuss the concept and experimental questions about the aether in great detail, even the null interferometer result for the static aether, as obtained by A. A. Michelson in the 1880s. Active questioning about the possibility of a dynamic (moving) aether was at the fore in the eleventh edition. The article closed in a very upbeat way, promising that much new was yet to be discovered about the aether. "These results constitute a far-reaching development of the modern or electrodynamic theory of the aether, of which the issue can hardly yet be foreseen." Yes, even electricity—previously an unknown "aetheric fluid" was becoming identified, in part, with the newly discovered electron. Matters of atomic transformation were just beginning to be divined.

It is 2003, and establishment science long ago threw out any serious discussion of the aether and its measurement. But the ghost of the aether is back. The spirit of Nikola Tesla lives, and there is much unfinished business at hand for physics. A sane, experimentally based cosmic view may yet be rescued from what masquerades as an increasingly "perfected" so-called Modern Physics. What did Tesla think about the aether? For that matter, what did Tesla think of "electricity"? We must remember that when the nineteenth-century Tesla worked, the aether was inextricably connected with the concept of electricity—in addition to its being the medium of the transmission of light and other Hertzian electromagnetic waves. The idea of "particles of electricity"—later to be discovered and then called "electrons"—was not yet in vogue. Electricity was thought of as something like an intangible fluid, literally "etheric." In a seminal talk before the American Institute of Electrical Engineers (AIEE) in May 1891 at what was then called Columbia College in New York City, Tesla spoke these telling words: "Of all the forms of nature's immeasurable, all-pervading energy, which ever and ever change and move, like a soul animates an innate universe, electricity and magnetism are perhaps the most fascinating . . . We know that electricity acts like an incompressible fluid; that there must be a constant quantity of it in nature; that it can neither be produced or destroyed . . . and that electricity and ether phenomena are identical." Tesla noted that this aether was everywhere moving and dynamic. The use of the aether would be the salvation of humankind, he said: ". . . with the power derived from it, with every form of energy obtained without effort, from stores forever inexhaustible, humanity will advance with giant strides." He said, "[I]t is a mere question of time when men will succeed in attaching their machinery to the very wheelwork of nature."

Of course, in Tesla's lifetime the wheelwork of nature—the aether—was not harnessed. It was to become very much out of style to be talking about an aether—any kind of aether, static or dynamic. The advent of Albert Einstein's relativity theory had begun to abolish the aether from the physicists' vocabulary in the 1920s and 1930s. Yet when *Time* magazine put Tesla on its cover in celebrating his seventy-fifth birthday (July 10, 1931), it referred to Tesla's work toward harnessing an "entirely new and unsuspected [energy] source." Was this from the aether? Perhaps.

Tesla had long hoped to be able to distribute electric power globally through the medium of the aether, power generated at the transmission point with benign and unlimited sources such as hydroelectric power. The power would be consumed only as needed at millions of receivers, being carried to each within the resonating cavity surrounding the body of the earth. Such power would be transmitted not by "electromagnetic radiation," as we ordinarily think of it (oscillating electric and magnetic waves transverse to the direction of propagation), but by longitudinal waves, which were more akin to longitudinal pressure waves in the air—the propagation of sound. He had conducted many experiments that seemed to show that such non-electromagnetic power propagation was possible. Indeed, Tesla illuminated electric bulbs at good distances. But was this really a new form of energy propagation? Indeed, it seems to have been.

Consider Nikola Tesla's special induction coils, which are called "Tesla coils" these days. Supposedly the only thing circulating in these coils or out of them is all that modern physics knows about or expects to be there—electrons for the "electricity" that can be in the wires of the coils, and "electromagnetic radiation" that can emanate from these coils. There can be no such thing as "longitudinal waves" emanating from such coils. Everyone knows that electromagnetic radiation is a transverse wave (from side to side, perpendicular to the direction of propagation), an electric and magnetic phenomenon in the nothingness of space-time, right?

For a long time, good experimenters have been puzzled about the workings of Tesla coils. It seems that Tesla coils are rich, indeed, with clues to the very structure of a dynamic aether. Recent experiments touch on the deep issue of the aether and its relation to what are evidently two basic forms of electricity, the accepted form (massbound, the flow of electrons), and the other not accepted at all by conventional science: massfree, capable of flowing in and around wires, as well as being transmitted as Tesla waves through gas media and vacuum (see scientific monographs at www.aethero metry. com). The massfree form of electricity might be called "cold electricity."

This hearkens back to another fundamental issue, the very nature of some nonstandard biological energies, which are also presumed not to exist and are the subject of much mockery these days. I have become convinced that these biophysical energies are integral to the nature of aether. When tracing back the origins of twentieth-century conceptions of organisms as purely biochemical systems, with nerve cell electric depolarization as the exclusive explanation for nonchemical, long-range signaling through an organism, one comes to the argument about "vitalism" or "animal electricity" that originated in the scientific controversy between L. Galvani and A. Volta in the late eighteenth century. It turns out that much was lost in the marginalizing of Galvani's "animal electricity" ideas of unipolar (single-wire) electric flow by the ascendant bipolar battery conceptions posed by Volta, which dominate our understanding of electricity today. But the conceptions of Tesla are in the process of returning in this, Tesla's third century. Tesla was very interested in the electrical component of life-energy, as was that other aether theorist, Lord Kelvin. Thus, we owe a lot to Tesla, not only for the technology that runs our world today, but also for the future energy sources that will abolish the Hydrocarbon Fuel Age and the future biomedicine that will seamlessly integrate Occidental medicine with the wisdom of the East. These matters can be adjudicated in laboratory experiment, even as the essentially psychotic scientific establishment—ignorant of the facts of its own history, and only too willing to insult some of its greatest benefactors—ignores it all.

11 Tom Bearden Fights for Revolutionary Science

A New Energy Pioneer Lays the Groundwork for Coming Discoveries

William P. Eigles

very revolution has its leading theorists, individuals who attempt to construct a logical, coherent formulation of new principles and concepts to rationalize and explain the occurrence of radical, paradigm-upsetting events or developments. Even if they are not there at the beginning of such seminal milestones, such individuals are quickly spawned in the aftermath, acting as compelling champions of the activists who are making the history. In the case of the revolution beginning to emerge more publicly in the field of alternative energy sources and technology, retired Army lieutenant colonel Thomas Bearden may soon be recognized as one of a small cadre of scientists and engineers who are just such credible boosters, convinced of and actively supportive of alternate energy realities early on.

Bearden recently delivered a paper on energy flow, collection, and dissipation in overunity electromagnetic devices at the International Symposium on New Energy in Denver, Colorado, where I had a chance to visit with him.

Big, bluff, and indefatigably ebullient in demeanor, Bearden first came to public notice in the early 1980s with the publication of his book *Excalibur Briefing*, in which he offered theoretical explanations for a wide array of paranormal phenomena and discussed various military applications of

Fig. 11.1. Thomas E. Bearden is a nuclear engineer, war games and weapons analyst, and military tactician. (Photograph courtesy of Tom Miller)

psychotronic devices powered by human psychic energy research in the United States and the Soviet Union. One of his many controversial claims was that the U.S. Navy nuclear submarine *Thresher,* which sank in the Atlantic Ocean with all hands on board in mid-1963, was the victim of an advanced operational Soviet psychotronic weapon. Since the early 1990s, however, Bearden has shunned any discussion of psychotronics, mysteriously claiming reticence to be the prudent course for any man interested in "staying healthy." This consideration also impels him to avoid any work on antigravity propulsion systems, which he became familiar with in his consulting work in the 1980s for the late inventor Floyd "Sparky" Sweet. It would seem that investigating certain areas of energy research, like the subject of government involvement with UFOs, entails more and greater risks, for undisclosed but perhaps easily inferable reasons relating to the nature of politico-economic power and those in our world who possess it in great concentration.

What Bearden is voluble about, however, and what occupies his time and attention almost exclusively these days, is his work on perfecting the theoretical scientific underpinnings of, and ultimately a verifiable model for, electromagnetic systems that legitimately produce more energy than they consume (known as "overunity" devices). Such systems propose to make use of the random electromagnetic fluctuations that exist in the vacuum of space, known variously as "free energy," "space energy," or "zero-point energy." Armed with an M.S. degree in nuclear engineering from the Georgia Institute of Technology and longtime employment in the aerospace industry, Bearden has researched this topic intensively for over twenty years, and currently serves as president of CTEC, Inc., his own research and development company, located in Huntsville, Alabama.

Bearden's work began with a reexamination of the fundamental concepts of classical electrodynamic theory, in light of the teachings of modern quantum mechanics and particle physics, in order to better understand how and why current actually flows in electrical circuits, where that energy comes from, and how it might be increased. This effort suggested to him major flaws in the paradigm established by nineteenth-century scientists James Clerk Maxwell and Hendrik Lorentz, whose equations and calculations (as they are known today) dealt only with the electrical energy that measurably flows in circuits and powers the devices that are attached. Analogizing to the water flowing around a fixed paddle wheel immersed in a river and the moving air surrounding a windmill, Bearden discovered that the free energy of space was knowingly ignored as a usable source of electrical energy by these scientists, and that the classical theory needed updating to reflect twentieth-century discoveries.

In Bearden's view, the principal faults in reasoning lay in two places. First, the algebra used to express Maxwell's original equations was changed, to ease understanding by others, from the highly complex quaternion type, which allowed and even prescribed overunity electromagnetic systems powered by space energy, to a much simpler tensor vector analysis, which did not. Second, Lorentz mathematically narrowed the scope and application of Maxwell's equations to describe only that part of the energy flow that physical circuits were designed to catch and use. In overall effect, according to Bearden, the early theorists made mistakes in interpreting their own calculations and unwittingly modified their original equations to discard a significant portion of the energy that was extractable—and, in fact, is extracted—from the vacuum by actual physical systems. The central issue for him therefore became: How does one redesign these systems in a new way to be able to collect and make effective use of this excess energy from the river of space, which demonstrably exists and is so readily available in the ambient environment? And then: How does one keep the redesigned systems from destroying themselves by over-tapping the infinite energy source of space?

Bearden has posited that an iterative collecting and scattering of space energy could be used to enable a quantum of energy to be reused multiple times, performing a quantum of work in each re-scattering. This iterative "retro-reflection" and multipass recollection would serve to increase the density of collected energy and therefore the local potential and strength of source dipoles (separations of charges) that occur in space due to the interaction between free charges and the vacuum. Bearden has labeled this process "asymmetric regauging" and believes that it increases the energy extraction by dipoles existing in the vacuum exchange. He believes that this process has been experimentally proved by the Patterson Power Cell, an innovative, recently marketed power device with a demonstrated overunity energy output.

Bearden's work progressed to clarifying the nature and characteristics of the two wave components of electromagnetic energy fields, the transverse waves and the longitudinal. Created simultaneously but traveling on different planes, Bearden likens the transverse wave to the easily perceived, slow waveform seen on the ocean's surface and the longitudinal wave to a swift-moving subsurface pressure wave that does not disturb the surface and is not capable of being measured by existing technology. Through the work of researchers Donnelly and Ziolkowski, Bearden found that, by science's current selection and use of the transverse wave to power conventional electrical devices, the hidden longitudinal wave is somehow "killed off," preventing it from being exploited to do useful work. The longitudinal wave, however, is potentially more powerful than the transverse wave, in that the former allegedly moves many times faster than the speed of light, which is a limiting factor for conventional signal transmission using the transverse wave component. Because of the theoretical ability of the internal longitudinal wave to facilitate virtually instantaneous signal communication across vast expanses of space, Bearden has focused on how to stimulate and select it for use, and allow the transverse wave to be canceled or not initially produced.

Bearden notes that he is now preparing a patent application for the initial part of what he terms "a superluminal communications system" that uses a longitudinal wave process and is capable of transmitting signals at speeds faster than the speed of light. He contends that the basic concept has already been shown theoretically and experimentally at a microscopic level by other researchers using waveguides. His team specifically intends to show how to form the longitudinal wave, by transmitting a video signal inside a DC voltage without any transverse wave signal accompanying it, and then retrieving the signal without the presence of any noise.

Bearden already has three patents now pending in the field of electric circuits, all of which purport to achieve overunity in energy output with absolute conformance to the conventional laws of physics. Nonetheless, he makes no claim to have yet developed a working model of any overunity device in his own laboratories. He does claim that his experimental results have been encouraging to date, and that as far back as 1990, his team was blowing up circuits due to the excess space energy they were tapping. The energy apparently could not be controlled in the semiconductor arrays being used at that time, which caused the energy to "ping-pong" back and forth between them until the resulting surge overloaded one of the arrays.

Bearden states, without disclosing more, that his team now knows how to control the energy flow, but is at a standstill for lack of funding. Fabrication difficulties have prevented forward movement on another means of energy flow control using a specialized, hard-to-engineer metallic material he has dubbed, with tongue in cheek, "Unattainium." However, he allows that his work in using multiple passes of energy, collecting it repeatedly using retro-reflection in electric circuits and thus enabling increased energy extraction, holds the most promise.

Bearden's work in this last vein may owe its stimulus to his consulting assistance to home inventor "Sparky" Sweet in the 1980s. Sweet had invented an assembly of wire coils and barium ferrite magnets that would extract energy from space and produce six watts of usable power, with only a much smaller trickle of energy as input. Dubbed a "vacuum triode amplifier" (VAT) by Bearden, a later model reportedly produced 500 watts of output power, showing a net gain of 1.5 million over the input power level. Bearden theorized that Sweet's device "tricked" the barium nuclei of the magnets into going into self-oscillation with the ambient vacuum, causing the fields of the specially conditioned "kinetic" magnets to quiver at a high level.

The theorist prevailed upon Sweet to make a change in his device that would allow for a test of antigravity properties. Sweet later reported to Bearden by phone that, by increasing the power output drawn from his device by adding greater loads, he was able to reduce the weight of the VAT, as measured by a scale, by 90 percent. Concern about the likelihood of exploding the magnets prevented Sweet from reducing the VAT's weight entirely and seeing it fly. Unfortunately, all of Sweet's secrets about how to activate his magnets to achieve his startling results died with him in 1995, and Bearden was left to pursue his theoretical research without the benefit of a working model.

The theorist and I discussed two of his books. One of them, *Energy from the Vacuum*, presents "the world's first legitimate theory of overunity electromagnetic engines, circuits, and devices," according to Bearden, and contains "a little necessary secret" essential to building them. The other book deals with Bearden's second and related main interest, the "Priore device," which was developed under the aegis of the French government in the 1960s and early 1970s.

Bearden reports that the Priore device is reputed to have cured terminal tumors in laboratory animals and is able to cure any disease, including arteriosclerosis and cancer, by a special electrodynamic process known as "phase conjugation" or "dedifferentiation." This process, seemingly miraculously, allegedly causes afflicted cells to return to their previously healthy state by literally turning back the clock on the disease. Bearden states that this process is a direct outgrowth of the work of the American Nobel Prize nominee Dr. Robert Becker, who demonstrated the use of small DC currents to heal intractable bone fractures by stimulating the growth of new bone. The trickle current apparently caused red corpuscles to shed their hemoglobin coating, grow new nuclei, and metamorphose into a much earlier, primitive version of the cells before differentiation. These cells could then be newly differentiated into needed bone cells, which would deposit themselves at the fracture point and result in a knitting of the broken bone. It is this basic process that Bearden asserts can be imported into the treatment of infectious and terminal diseases, including the restoration to health of the immune systems of people with AIDS. And, Bearden claims, the Priore mechanism can effectuate healing in a matter of minutes.

Looking to the future of both Priore devices and overunity electromagnetic systems, Bearden sees the greatest obstacle to their realization being the mind-set of the existing research-funding establishment and the orthodox scientific community that it serves. The flow of funding effectively controls what research is pursued by scientists working at universities and in industry. The mind-set against the possibility of tapping and collecting space energy to provide usable electricity serves to block the allocation of money to develop working prototypes. The early new energy pioneers who have most influenced Bearden in his own efforts, Nikola Tesla and T. Henry Moray, faced this same mind-set, resulting in their work being ignored by the scientific community of their time and eventually being suppressed by various contemporary interests.

Still, Bearden remains optimistic. He believes that once a scientifically verifiable model is perfected that is consistent with modern particle physics and thermodynamics and working, experimental proof is clearly established— thereby dispelling any notions that perpetual motion is being proposed—the mainstream scientific community will begin to lend support and the race to a new energy future will be on in a big way. He foresees commercially marketed overunity devices becoming available in two years, with homes and cars later being powered by insertable solid-state, energy-collecting cards. And, with the advent of the Internet, the ubiquitous availability of modern communications links, and the proliferation of journals and newsletters dedicated to alternate energy technology, the ability of a hostile establishment to suppress scientific innovation and its proponents is now greatly reduced. The new-energy genie, once out, will be much tougher to get back into the bottle than in earlier decades.

For his part, Bearden believes that his major contribution will be to "have blown a hole in the brick wall, not a nice door," of the traditional way of thinking about overunity systems, primarily as a theorist rather than an inventor. He expects that interested, bright graduate students and postdoctoral fellows will take matters to the next level. Only time will tell.

Although Bearden is not without his detractors, he is an undeniably engaging and colorful character whose deep conviction about his work and its results inspires both fascination and curiosity. If, in conversing with him, you were to evince any doubt about his claims, Bearden is quick to point out, "This is not Tom Bearden [talking], it's in the [scientific] literature! If only people would read it and test it." Agree with him or not, he is, at the very least, a visionary of almost evangelical fervor who is sincerely dedicated to helping develop a new source of usable energy that is cleaner, cheaper, safer for the earth and its peoples, and universally available worldwide. To be sure, that's a goal worthy of everybody's attention.

12 **Sonofusion**

Can the Energy of the Sun Be Captured in a Bottle—Like an Ancient Genie—and Provide Us with a Future of Limitless Energy?

John Kettler

What do nail polish remover and your ultrasonic toothbrush have in common with a possible fusion breakthrough? More than you might think! In fact, those two things, acetone and ultrasonics, may well prove to be the means by which fusion research no longer requires multibillion-dollar Tokamak magnetic containment vessels and a many-armed high-energy laser bearing the name of the Indian god Shiva. Potentially fusion—the same process that fuels the sun—could be done on a tabletop.

The other good news here is that if the principal experiments are successfully replicated by other scientists and the new technique begins to find favor in nuclear research circles, then—unlike cold fusion, which sprang from chemists unsanctioned by the physicists—sonofusion may actually thrive, for it comes from the hot fusion physics community. Indeed, the first two names on the seminal sonofusion paper in the March 8, 2002, issue of *Science,* "Evidence for Nuclear Emissions During Acoustic Cavitation," R. P. Taleyarkhan and C. D. West, are both associated with Oak Ridge, Tennessee, one of the cradles of nuclear research in the United States. R. P. Taleyarkhan is at Oak Ridge National Laboratories, while C. D. West is with the Oak Ridge Associated Universities. This might seem a kind of imprimatur and ward against the kind of attacks that greeted cold fusion in 1989, but that definitely wasn't the case. The paper very nearly didn't get published at all. More on this later.

SONOFUSION DEFINED

Sonofusion is a neologism meaning "fusion from sound." That's good as far as it goes, but it's only part of the story. Sonofusion is a kind of outgrowth from an emerging chemistry field called sonochemistry, and particularly a phenomenon called sonoluminescence. Yes, we now have two more coined words, but we're closing in on our quarry.

Sonoluminescence occurs under certain conditions of temperature and pressure when appropriate fluids, such as acetone, are strongly pulsed by ultrasonic waves. These waves cause myriad tiny bubbles to form, much like the ones created by your toothbrush but of vastly greater energetic content. When these bubbles collapse, part of their energy is sometimes released as light—hence the phenomenon's name, sonoluminescence.

The idea for sonofusion is the apparent result of observing the enormous pressure and temperature in the bubbles undergoing sonoluminescence and testing to see what would happen if one were to "salt" the medium with the fusionable element deuterium, a hydrogen isotope, then really agitate the already stressed mixture by injecting high-energy neutrons. This, in a nutshell, is what Taleyarkhan et al. claim they did, and the work was controversial before, during, and after publication.

SCIENCE: THEORY VS. PRACTICE

Science loves to depict itself as being an ongoing, open-ended, freewheeling inquiry into the nature, structure, function, and interactions of things ranging from the infinitesimally small to the incomprehensibly large, from the simplest to the most complex, bound only by the canons of integrity, experiment, peer review, and reproducibility of experimental results. This is how it's supposed to work in the standard "noble scientist" depictions.

Unfortunately, the world of science more nearly resembles a cross between a Super Max prison and an insane asylum, with strong elements of the Inquisition clearly in evidence.

At any given time, science is defined by a handful of dogmatic views and, rest assured, enforcers to protect said dogma abound at all levels. Schools teach only certain versions of reality, be they Flat Earth, Ptolemaic astronomy, or Darwinism, to name but three down through the ages, and heaven help anyone who comes up with something that threatens the standard model, for now egos become involved, academic positions of several sorts are threatened, and frequently external secular and religious authorities become involved, sometimes with fatal results. Just ask Giordano Bruno!

In modern practice, science is the handmaiden of the state and business and depends upon them for funding and other types of support. They call the shots and science dances to the oft incestuous tune, for the work is capital intensive, sometimes involving facilities exceeding billions of dollars, pricey academic and engineering talent, and loads of expensive lab equipment and supplies.

The state has by far the bigger input, not only in that it employs hordes of scientists in myriad organizations and laboratories but also in that it funds basic research as well as research and development by think tanks and industry. The upshot of all this consists of lots of tame scientists instead of noble untrammeled explorers. Their job is to not upset anyone, make the boss look good, protect their organizations from legal or other damage, grow their careers, and expand their organizations. For many, it's now also about securing lucrative patents from taxpayer-supported research. Such a climate does not exactly foster open investigations and the free sharing of results. It grows worse when considerations of security classification, business confidentiality, patent protection, and the like are thrown in.

Scientists who buck the system or pose a real problem are subject to all kinds of controls. They can be denied permission and funds to attend professional conferences, be denied advancement, be blocked from publishing, be informally blackballed, be expelled from scientific societies, censured, fired, or even worse. Some maverick scientists have been threatened or assaulted, endured repeated break-ins and lab destructions, and even been killed. Some have been jailed, and others have had their writings and equipment seized and destroyed.

The more fundamental the discovery and the more vital the interests that are threatened, the worse the back blast becomes, and in this clash the media, scientific and mass, figure prominently. It is here that reputations and scientific battles are often won or lost, which is why controlling who publishes what and where is so important. This was certainly true for "Evidence of Nuclear Emissions During Acoustic Cavitation."

SCIENTIFIC PUBLISHING—A SHORT COURSE

The way it's supposed to work is that a scientist or group thereof, sometimes scattered around the world, decides to test a particular hypothesis and carefully crafts a controlled experiment designed to prove or disprove that hypothesis. That experiment is then meticulously documented, with the results and arguments pro and con being presented, if something novel is found, in a formal scientific paper, which typically has to be staffed through multiple internal bureaucracies before ever leaving the premises.

The paper is then submitted to the editor of a peer-reviewed scientific journal appropriate to the purported discovery. The editor then sends out the paper to a panel of suitable scientists, with the composition of the panel usually being anonymous to the public. It is the job of this panel of peer reviewers to scrutinize the paper to see that it is scientifically solid work, with a properly structured and documented experiment, and all identifiable sources of bias and error accounted for. Comments from the peer reviewers go back to the editor, who then has to decide whether to reject the paper outright, obtain clarification on certain points from the authors, or publish the paper as written, subject only to standard editing practices. Once a paper is accepted for publication, it's published and that's that. So goes the theory.

THE PAPER WITH MANY LIVES

The above certainly wasn't the case with this paper on sonofusion, for there was interference from many quarters, with the evident intent of preventing the paper's publication or, in some cases, of at least diminishing its impact and public profile. And this was for a paper that had passed its peer review and was scheduled for publication.

The situation was so bad that *Science* editor Donald Kennedy took the highly unusual step of describing the machinations and pressure in his editorial "To Publish or Not to Publish," which accompanied the issue of *Science* in which the paper appeared. First, the senior science managers at Oak Ridge National Laboratory got cold feet, belatedly expressing reservations over both the findings and their meaning and repeatedly asked for a delay. This triggered both meetings and negotiations in which findings from a second set of scientists (D. Shapira and M. Saltmarsh, both at Oak Ridge) were also addressed and some were incorporated. Further, the paper's authors modified their original text and also cited a non-peer-reviewed message from the second group and responded to criticisms raised therein—highly unorthodox procedures, to say the least.

Nor was Donald Kennedy pleased by the incident. In yet a third piece in the same issue of *Science,* Charles Seife's "'Bubble Fusion' Paper Generates a Tempest in a Beaker," Mr. Kennedy is quoted as saying, "There was certainly pressure from Oak Ridge to delay, if not kill, the paper." He went on, "I'm annoyed at the intervention, and I'm annoyed at the assumption that non-authors had the authority to tell us we couldn't publish the paper."

With the above already in train, *Science* got letters from physicist William Happer of Princeton and Richard Garwin of IBM's Thomas J. Watson research laboratory, gentlemen not named in the editorial but characterized by the editor as being "distinguished scientists in this field." Their letters raised objections and urged that publication be reconsidered. The "Bubble

Fusion" article identifies Happer as the former head of the Department of Energy's science office for two years in the early 1990s and says that his concerns were twofold: preventing *Science* magazine from shooting itself in the foot and saving the scientific community from another public humiliation— "I saw it happen with cold fusion. If we're really unlucky, Dan Rather will talk about it on the evening news and intone how, providentially, the energy problem has been solved. We as a community will look stupid."

This is precisely the kind of advancement-suppressing thinking and bureaucratic risk-averse self-protectiveness previously identified as separating the popular concept of science from how science is actually practiced.

Richard Garwin at least raised a substantive issue in his letter, expressing concern over how constant tweaking of the equipment necessary to maintain proper test conditions may have allowed unconscious bias to occur in the results. Given the above, Mr. Seife quotes him as saying, " . . . it would be unfortunate if *Science* magazine were to take any position on its correctness."

Additional pressure was externally applied by Robert Park, of the American Physical Society, who, before the paper was even published, airily dismissed it.

Donald Kennedy isn't sure whether or not the paper's results are reproducible, but that isn't his job. As he noted in his editorial, " . . . our mission is to put interesting, potentially important science into public view after ensuring its quality as best we possibly can . . . efforts at repetition and reinterpretation can take place out in the open. That's where it [*sic*] belongs, not in an alternative universe in which anonymity prevails, rumor leaks out, and facts stay inside."

From the above, it should now be clear to the readers that only two parties here played the publication game fairly and by the book: the authors of the paper and the editor of *Science* magazine, Donald Kennedy, clearly a man of courage and integrity.

Let us now revisit our admittedly oversimplified model of the experiment and flesh it out some.

SONOFUSION: FLESHING OUT THE EXPERIMENT

Imagine two identical small chambers, each filled with carefully degassed nail polish remover (acetone), one of which has had deuterium added to it by replacing the hydrogen atoms normally found in acetone. Both chambers are put under a vacuum, brought to 0 degrees Celsius (32 degrees Fahrenheit), then hit with both powerful acoustic pulses from an ultrasonic generator while also being bombarded by a 14 million electron volt (MeV) neutron source, giving rise to myriad nanometer-sized bubbles, which soon grow to such huge proportions that they can be seen (about a millimeter across). Their collapse is detected by a microphone in contact with the chamber walls, and any sonoluminescence is picked up by a photomultiplier tube, a device capable of amplifying small amounts of light to levels perceivable by the human eye and displaying what's detected. The other major piece of apparatus for both chambers is a scintillation detector, used to detect nuclear events.

Taleyarkhan and colleagues claim in their paper that something amazing, to us at least and many scientists, happened when the deuterated acetone was bombarded by ultrasound and neutrons. What exactly? Tritium was formed, and 2.45 MeV neutrons were detected. If true, these are solid proofs that nuclear fusion, deuterium–deuterium, occurred under extreme pressures and temperatures as the bubbles imploded and sonoluminesced.

Interestingly, the reported observations cover both ends of two equally likely outcomes in a deuterium–deuterium fusion reaction. One produces helium-3 and a 2.45 MeV neutron, while the other produces tritium and an absorbed-by-the-acetone 3.02 MeV proton. No such results were found in the test chamber filled with normal acetone, nor did they occur in the deuterated (deuterium-enriched) acetone solution when the temperature

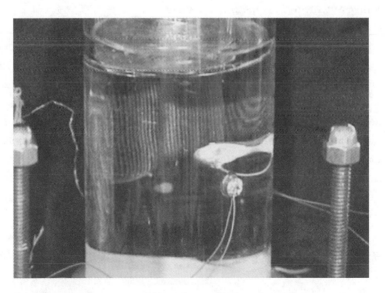

Fig. 12.1. Apparatus used in early sonofusion experiments.
(Photograph courtesy of the Oak Ridge National Laboratory)

was changed to impede cavitation or when the ultrasonic generator was switched off.

WHAT NEXT?

Now that the paper has been published, in spite of gross interference and multiple attempts to kill it, it falls upon other scientists in the scientific community to set up and run their own experiments. Some wish they'd thought of doing the experiment themselves; others concede that fusion could've occurred as stated but say that proof is lacking; still others argue that this or that factor wasn't accounted for, that contaminants were in the tritium, or that the neutron count is off. It gets even more abstruse after that level.

But this is how it's supposed to be, for the idea is that only by thoroughly going through a supposed experimental breakthrough and by running independent experiments to confirm or deny the initial one will the truth be arrived at. Of course, this presumes that people are playing fairly. Sadly, it is not at all unheard of for cheating to occur, for apparatus to be deliberately misconfigured, test conditions to be changed, and results outright falsified. Character assassination is also frequently used. All of the above nastiness must be carefully screened for in looking at attempts to independently replicate the experiment. This is particularly true when the scientists doing the follow-on work belong to an entity with big economic interests at stake via heavy investment in a competing line of work.

Organizational behavior teaches us that the first priority of an organization is survival, and there is a natural human tendency to reject the unfamiliar and to protect one's income and social position. The hot fusion community has been suckling at the state's teat for decades to the tune of untold billions of dollars, and it has powerful, politically connected suppliers that provide its exotic equipment. Think of the various empires, satrapies, and fiefdoms at risk in government and business if sonofusion really works, if fusion really can be done on a tabletop.

Are sonofusion power plants around the corner? Are we on the verge of being able to extract practically unlimited thermonuclear fuel from the sea? Hardly. R. P. Taleyarkhan is quoted in PhysicsToday.org, in "Skepticism Greets Claim of Bubble Fusion," as saying, "We are nowhere near power generation."

13 Escape from Gravity

Is the Ancient Promise of Freedom from the Pull of Earth Alive and Well at NASA and Elsewhere?

Jeane Manning

Do we live in a magical era at the same time as we are experiencing stressful days? I find that awe and wonder are very much alive, as paradigms (worldviews) are being stretched. Individuals say their expansion comes not only from spiritual experiences, but also from exposure to scientific breakthroughs.

When they see the "impossible" becoming possible, it seems, people are thinking more about the primal sea of energy that supports such marvels. Consider two small examples of paradigm expanders—a tiny levitating toy in the United States and a perpetual-motion sculpture in Norway. And in the larger physical arena, according to visionary engineers, our collective worldview is expanding because we are closer to starship travel than most people believe.

Starships? Yes, antigravity technologies may literally get off the ground in the near future. Scientists are taking seriously the possibility of an "inertialess drive" for spaceships.

OVERCOMING INERTIA = LIFTOFF

Inertia is the tendency of objects in motion to keep on moving in the same direction and of a body at rest to remain on the sofa. When you are standing on a bus that starts with a jerk or stops suddenly, inertia is the force that throws you onto the floor. Then there are the "g-forces" that contort the faces of people in an accelerating rocket.

Gravity and inertia must be overcome somehow if spacecraft are to perform tricks attributed to supposedly extraterrestrial objects in the sky. Viewers, including airline pilots, have described unidentified craft that make sudden sharp turns without reducing speed or that accelerate from hovering to high speed. For occupants of a spacecraft to survive the sudden changes of location, inertia would have to be canceled or manipulated in and around the object. This would be in effect a controllable gravity field.

The possibility of "inertialess drive" is nearer to us, because mainstream scientists now have a picture of what might be the cause of inertia. A few years ago the respected physics journal *Physical Review* published a paper by B. Haisch, A. Rueda, and H. E. Puthoff with a theory about inertia. They point to the fact that what is popularly known as empty space is not empty; throughout the universe it is seething with "zero-point quantum fluctuations" of electromagnetic energy. The three physicists suggest that interaction with this zero-point field causes both inertia and gravity.

If we understand that interaction, can we go to the stars? Maybe understanding it is a first step. More recently, one of those three physicists—Dr. Hal Puthoff—elaborated. In the science magazine *Ad Astra,* he writes about the "vacuum" of space as an energy reservoir, with energy densities as powerful as nuclear energy or greater. If the zero-point field (ZPF) could be "mined" for practical use, it would, everywhere in all galaxies, supply energy for space propulsion.

Fig. 13.1. A yogi demonstrates levitation in the June 6, 1936, Illustrated London News.

How would it work? Puthoff gives clues—such as a phenomenon called the Casimir effect, which pulls together closely spaced smooth metal plates. Another researcher, Robert Forward, has demonstrated how electrical energy could be taken from the electromagnetic fluctuations of the vacuum by manipulating this effect. Puthoff also cites a paper by his coauthors, Haisch of Lockheed and Rueda of California State University, along with Dr. Daniel Cole of IBM: "They propose that the vast reaches of outer space constitute an ideal environment for ZPF acceleration of nuclei and thus provide a mechanism for 'powering up' cosmic rays." He mentions a report published by the U.S. Air Force about the possibility of using a "sub-cosmic ray" approach to "accelerate protons in a cryogenically cooled, collision-free vacuum trap and thus extract energy from the vacuum fluctuations . . ."

What it boils down to, Puthoff says, is that scientific experiments indicate that human technology can alter vacuum fluctuations. This leads to the related idea that, in principle, we could also change gravitational and inertial masses.

Puthoff points out that accepted theories up until now looked only at the effects of gravity and inertia instead of at the origins of these fundamental forces. He notes that the first scientist to hint that gravity and inertia might be rooted in the underlying "vacuum fluctuations" was the Russian dissident Andrei Sakharov in a 1967 study.

Concluding his *Ad Astra* article with a quote from science-fiction author Arthur C. Clarke saying that highly advanced technology is indistinguishable from magic, Puthoff adds that "fortunately such magic appears to be waiting in the wings of our deepening understanding of the quantum universe in which we live."

FROM SCI-FI TO NASA'S NEW TEAM

Arthur C. Clarke honors Sakharov, Haisch, Rueda, and Puthoff in his latest novel, *3001: The Final Odyssey*. Clarke names his fictional inertia-canceling space drive SHARP, an acronym made of the four scientists' names, and *3001* cites their *Physical Review* paper as a landmark. In a commentary at the back of his novel, Clarke notes that controlling inertia could lead to interesting situations. For example, "If you gave someone the gentlest touch, they would promptly disappear at thousands of kilometers an hour until they bounced off the other side of the room a fraction of a millisecond later."

It takes a sci-fi writer to present the extremes of what could be possible. On the other hand, institutional science is closer to near-future prospects for technological change. One institution, the National Aeronautics and Space Administration (NASA), is assembling a team for a breakthrough propulsion laboratory, sparked by Marc G. Millis. Millis, of the Space Propulsion Technical Division, NASA Lewis Research Center in Cleveland, last year wrote a technical paper about the new theories that suggest that gravitational and inertial forces are caused by interactions with the electromagnetic fluctuations of the vacuum.

So it now appears respectable to say that antigravity devices are possible, and that they could conceivably operate by manipulating the free energy in space—otherwise known as ether, or aether.

"There have also been studies suggesting experimental results for mass-altering effects and a theory suggesting a 'warp drive,'" Millis says. "With the emergence of such new possibilities, it may be time to revisit the notion of creating the visionary 'space drive.'" A "space drive" would propel the perfect starship—it could use the fundamental properties of matter and space-time in order to whisk itself to anywhere in space without having to carry and expel an explosive fuel.

Will sci-fi travel to our neighboring stars become a reality in our time? Breakthroughs in science are needed, say Puthoff, Millis, and others—breakthroughs such as self-contained propulsion that needs no propellant. To chart a path toward such discoveries, Millis imagines different types of space drives. His hypothetical drives show the unsolved challenges that NASA must meet, and his paper breaks up the problem into research goals such as:

- Discover a way to "asymmetrically interact with the electromagnetic fluctuations of the vacuum."

- Develop a physics that describes inertia, gravity, or the properties of space-time electromagnetically. This research could lead to using electromagnetic propulsion technology instead of burning fuel.

- Find out if negative mass exists or if its properties can be synthesized. If negative mass doesn't exist, then a goal would be to have another look at concepts of the ether or other physics principles.

The hypothetical space or field drives Millis describes, such as "diametric drive, pitch drive, bias drive, disjunction drive," and his "collision sails" are too technical for this article. The point, however, is that times have changed. Previously, prospects for creating a space drive seemed too far in the future to justify NASA's creating jobs in such research. But now we have a team approach to the "space drive" challenge.

The vision is clouded elsewhere, however, by academics' fear of being associated with controversial concepts such as antigravity. Recently a scientist in Finland brought an unwanted flood of reporters to his university when his work—finding that spinning a superconducting ceramic under an electromagnetic field caused materials suspended above it to lose weight—was published in the *London Times*. The situation has parallels to the experience of Drs. Stanley Pons and Martin Fleischmann when they announced table-top "cold fusion" in Utah in 1989 and were hounded out of the country.

The materials scientist in Finland, Eugene Podkletnov, formerly with Tampere University, blames the popular press for sensationalizing his work and therefore ruining his project. But he recently told magazine writer Rob Irving that in five or seven years the antigravity effect could have been used to help replace polluting jet airplanes.

OUTSIDERS CLAIM ANTIGRAVITY

During the last half of the twentieth century, individuals working without institutional funding have also apparently made antigravity breakthroughs. David Hamel of Canada, the late T. Townsend Brown of the United States, and John Searl of England are among them. Brown was university educated and at times had one foot in the military's door, but Hamel and Searl are completely outside of the mainstream. The two have much in common.

For the first edition of *Atlantis Rising*, David Lewis wrote an article, "Is Anti-Gravity in Your Future?" about the unpretentious John R. R. Searl. The British inventor "appears to be a simple, honest man, gifted with an earnestness and a grasp of science that transcends his presentation . . . his grammar and speech are poor, his accent thick . . . no glib or polished spokesman which, ironically, adds to his credibility." Lewis remained prudently skeptical, but related the Searl anecdotes.

Searl invented what he called the Searl Effect Generator (SEG), which he claims has powered flying disks. Not little Frisbees, but models that could make a serious dent in any airplane they might accidentally ram. However, he didn't set out to make anything fly; he wanted only to produce power. In 1952, Searl is said to have built a fourteen-foot spinning SEG that created

Fig. 13.2. A version of inventor John Searl's flying disk under construction.

Fig. 13.3. Purported photograph of Searl's disk in flight.

an exceptionally high voltage. Instead of braking its speed, the SEG accelerated, ionizing the air around it. It broke away from any connection with the ground and disappeared into the sky.

Remember, this allegedly happened in the 1950s. Why didn't university professors and other government-funded researchers investigate Searl's claims? Part of the answer is that they would have been ridiculed. Even as recently as twenty years ago, NASA, for example, was less open than it now is to the nonconventional ideas. In the 1970s a NASA consultant, the late Dr. Rolf Schaffranke, was forced to write under a pseudonym, Rho Sigma. The result was a small book entitled *Ether Technology*. It told the story of Searl, along with that of T. Townsend Brown, whose experiments also pointed toward space flight without stress and without pollution.

A decade after Schaffranke's book came out, in 1989, he was in Einsiedeln, Switzerland, when nine hundred engineers met for the Swiss Association for Free Energy conference. One advertised attraction that brought me to that conference was the chance to hear the legendary John Searl. In a packed side room the night Searl arrived, the beleaguered inventor, now in his sixties, was welcomed warmly. He responded with an emotional outpouring of his story of hardships and harassments including a disastrous fire that burned both his equipment and his own skin. He publicly vowed that noth-

Fig. 13.4. Maverick British inventor John Searl. (Photograph by Tom Miller)

ing could stop him now. However, it is not easy to pick up the pieces of a tattered dream. Whether or not he is able to re-create his flying disks, perhaps others will build on Searl's inspiration.

POWER TO THE PEOPLE

Searl's story parallels David Hamel's in many details. Hamel is self-described as a simple man, a carpenter, who left his formal education behind after the fifth grade. For the past two decades he has built experimental devices that incorporate magnets with unexpected results, such as an experiment exploding through the roof of a garage workshop. Later Hamel built an eight-foot model of his machine on top of scaffolding in his front yard. When he set it into motion one night, it created a colorful aura of ionization around itself. To Hamel's dismay it rose into the air, soared higher and higher, and sped out of sight in the direction of the Pacific Ocean.

Instead of rejoicing about this proof that he could build a flying device, Hamel was dismayed that his investment in magnets had flown away. He is now building a large model out of polished granite and other heavy materials. When it is completed, Hamel will make certain that he has a large audience to watch the test run. His colleague on the west coast of Canada, electronics specialist Pierre Sinclaire, is equally determined to liberate us from outdated technologies. Sinclaire is selling a video for technicians who want to build the Hamel device, and will use the proceeds toward the cost of completing his own model by next year. After these new approaches to generating electricity and new transportation possibilities are proved, they intend to give their knowledge to the people of the world, and not merely to a select group.

ARE WE READY FOR LIFTOFF?

Also intent on empowering people, marketers of an antigravity toy are taking the public education route, and having fun while spreading awe and wonder. The Levitron is a top that achieves sustained levitation while powered only by permanent magnets. One marketer includes a flight kit with an instructional video on "the art of levitation." Mike Stewart, of New Mexico, tells me that more recent "magical" developments to educate people include a "perpetuator," which pulses the Levitron and keeps it going indefinitely. "The whole understanding of magnetism is up for grabs," Stewart says.

Watching this top hovering above its base for five minutes, I see clean energy and antigravity possibilities for our future expand before my eyes.

As for the question "Can we go to the stars?" perhaps we should phrase it as in the children's game "Mother, May I . . .": "Mother Earth, may we . . . ?" If there were an entity embodying Mom Earth, I think she would reply, "You can go and play among the stars after you've cleaned up your mess here at home!"

Mess, what mess? We could ask one of the scientists I met at conferences where he was a speaker in the 1980s (including the above-mentioned one in Einsiedeln, Switzerland). Adam Trombly founded an information network called Project Earth. Always he spoke eloquently about the state of the planet and provided abundant facts to back up his urgent messages. Trombly's message is now on the Internet at www.projectearth.com.

He also invented a nonconventional generator to harness the zero-point fluctuations of the "vacuum" of space. I view his technology as yet another star on the horizon beckoning us toward an exciting future. By first learning how to work in harmony with nature on our own planet, in my opinion, we could prove that we're mature enough to handle these advanced technologies responsibly, and then—let's go!

14 Power from the Nightside

Could Earth Itself Be Trying to Provide the
Clean Abundant Energy We Need?

Susan B. Martinez, Ph.D.

Mobil and Exxon won't like it. Nor will the nuclear power industry. Even solar power enthusiasts may frown in disbelief. But give them time. All new ideas, especially those that reverse conventional wisdom (180 degrees exactly), fall on deaf ears, unwilling ears.

Take polar energy, for example: that glorious spectacle of colored lights painting the night sky of the northern and southern latitudes. Auroral readings taken in the Van Allen Belt are on the order of three million megawatts. That's four times the power used in the United States at peak (summer) demand!

Can atmospheric conditions be used as power? Some think so.

Alaskans have begun to investigate the possibility of harnessing energy from those stunning nocturnal displays—known to us as aurora borealis, to the Maoris as Burning-of-the-sky, and to Europeans as Merry Dancers— shimmering, swaying, and waltzing across the firmament with dazzling grace and ineffable beauty. Yet, chances of tapping this almost occult power are slim, barring a revised (really, reversed) view of terrestrial mechanics: we need to know where auroras come from before launching the intriguing business of capturing their energy for the use of humankind. And if Ray Palmer was right, those flickering fireworks of the polar skies originate, not from the heavens above, but from the very bowels of the earth. "Our recent ISIS satellite," commented Palmer, a founding editor of *Fate* magazine, "has just [ca. 1970] confirmed . . . that the energy that causes the northern lights flows upward from the North Pole, rather than downward from outer space (from the sun) as previously held by scientists." Ray's "previously" was optimistic, for today's science stronghold, stubbornly ignoring its own findings, would have us believe that our cascading, gliding auroras are triggered by the far-off sun. And here's the explanation, the "sciencespeak," that makes it so— entrapment: solar particles that "get trapped in our planet's magnetic field" bump into atmospheric gases, causing them to glow.

But can it be?—given the perfectly well-known fact that gases of earth's atmosphere occur only in molecular form, while auroral wavelengths are frequently atomic: that is, atomic nitrogen, atomic oxygen, and so on. Where does this atomic energy come from if, as scientists have observed, ordinary energy from the sun cannot radiate these lines—that is, auroral rays?

With science to the rescue, a new kind of "high energy particle" (from the sun) is quickly postulated to fit the bill. And the explanation becomes: "solar wind." Conveniently, the fancied solar wind (invented in 1958) is also assigned the difficult task of pushing the auroras toward the earth's polar regions (missing all other latitudes), thus killing two birds with one stone, for how else may we account for the fact that a "donut" of light (auroral ring) favors the poles, as pictured by satellite images over the Arctic and Antarctica? And why would the solar wind perform this prodigy only at night?

Let's just stand this thing on its head and see how right Ray Palmer may have been. Let's change the direction of this auroral energy and suffer it to emerge, nightly, from the center of the earth. This powerful current, the earth's own motor or dynamo, thus completes its round-trip by emerging from "the northern pole . . . in flames of fire, which are called borealis."

Picture it: the earth-body breathes in by day and out by night, the north pole, nicely dented (no one knows why), serving as the primary vent, the chute, for earth's powerful effluvium.

Careful observers, like the American physicist William Corliss, have quietly admitted that "some auroras may record the slow discharge of terrestrial electricity to the upper atmosphere." And so it is . . . away from the earth, to the atmosphere, gushing, shooting outward and loaded with bolts of free and clean energy.

Such currents, Corliss goes on to observe, "that create auroral displays are accompanied by similar currents in the earth's crust." In fact, the same researcher is struck by "puzzling observations linking auroras to . . . earthquakes and mountaintop glows." But these observations are not so puzzling once the terrestrial origin of the Merry Dancers is allowed and, too, once the various "vents" (secondary vents) are identified.

The poles are by no means the only path of escape for the boundless energy that, after sunset, surges through the heart of the planet on its way back to the atmospheric dynamo (vortex) from whence it came.* The Brown

---

*The force of the vortex (roughly, geomagnetic field) is toward its own center, but turns at the center and escapes outward at the north pole—as one may draw a line from the east to the center of the earth, then in a right angle due north, which would be the current of the vortex.

Mountain Lights of North Carolina, for example, rise brilliantly over the ridge like "a bursting skyrocket," especially on dark nights. These lights, like auroras with their famous split-second changes, "suddenly wink out." The phenomenon generally known as the Andes glow—occurring also in the Alps, Rockies, and so on—involves bright flashes of colored light emitted from mountain peaks and shot high into the sky at great speed, sometimes seen hundreds of miles off. It is worth noting that the Brown Mountain area is seismically active, just as British studies show "a clear connection between these light displays and the presence of fault lines"—giving us our "vents."

In fact, there is a striking family resemblance not only between auroras and Andes glow, but also with earthquake lights, airglow (the inexplicable ambient light in the sky at nighttime), marine phosphorescence, volcanic glow, spontaneous earth fires, and marsh lights such as spooklights and will-o'-the-wisps.

Ruptures deep in the earth give us volcanoes and earthquakes, the latter—in Chile, Japan, China, and California—at times filling the sky with "earthquake lights" discharged from the depths. Just before the great San Francisco quake of 1906, tall blue flames played over foothills and marshland. Wastelands, swamps, and even graveyards produce phenomena intriguingly similar to auroras. Yet the most prevalent of marshland lights—will-o'-the-wisps—are so strange and mysterious that "serious scientific studies are non-existent."

These playful phosphorescent flames, small armies of which magically erupt at night over swampy land, command the selfsame descriptions as the floating auroras. The "soft eerie light" of will-o'-the-wisps compares readily with the ethereal curtains and supple draperies of the northern lights. The "ghostlike quality" of the swampland flames matches the "ghostly veils" of auroras. If will-o'-the-wisps prance and hop, auroras are Merry Dancers. If will-o'-the-wisps change colors instantaneously and disappear in a flash ("like a cinder"), auroras do the same. Appearing just a few feet off the ground, these small but bright pyrotechnics of the marshlands are nocturnal only, and they are of the earth—like Andes glow, spooklights, "money lights,"* "fairy lights,"† airglow, ginseng glow, and assorted terraqueous

---

Luces del dinero (lights of money) can be seen hovering over the land in Mexico and the Peruvian Andes. Their name derives from a superstition that if you see these lights, you will find a treasure.

†In Australia, fairy lights are also known as "Min Min Lights." Twinkling lights that appear in the desert, their source of illumination is unknown.

luminosities—all of which manifest only at night. What is that "special earth vitality" that, according to the Chinese, makes ginseng glow at night? (That's how hunters find it.) And what exactly is the energy that enables vegetable growth to take place mostly at night?

We could have none of these "marvels" if the earth itself did not possess a specifically nocturnal energy as yet undreamed of by human science, the cabal that would repudiate the well-known ghost lights of various locales, attributing them to some dull, prosaic cause, such as refraction from head-lights (even though the phenomena predate the invention of automobiles), or blaming them on natural gas (even when the marsh flames burn cold), or invoking that old standby—collective hallucination! Yet the enigmatic earthlights are real, and their spectral quality faultlessly imitates the gliding auroras—mobile and shape-shifting, "blinding" (Esperanza Light), "danc-ing in the dark" (Marfa Lights of Texas), color-changing (Summerville Light of South Carolina), and "blinking out" (Ozark Spooklight, near Joplin, Missouri). Arising from the ground, our ghost lights, witnessed constantly by thousands, speak the language of earth's effluvium in their every tour de force. Like Puck, the British hobgoblin, these spectacular visions emerge only at night. They are not freaks of nature; they are part of nature, part of the same grand plan that gives us day and night, and they are endowed with all the idiosyncrasies of the aurora borealis, which are spewed in lavish array from the planet's primary vents—the poles.

Everything that spouts from the nightside earth is marked by the same implacable force that thrusts auroras high in the sky, like volcanic ash that is propelled twenty miles into the stratosphere. We recognize in these pro-digious earthlights the familiar pigments—greens, yellows, blues, reds—of the colorful auroras. The sulfur/ozone smell of auroras is also perfectly analogous to that of waterspouts, earthquakes, volcanoes, surprising beach flames, and mud fires. Too, the vast and wondrous rotating wheels of light, perfectly geometrical and witnessed by astonished observers on the seas, are identical to auroral wheels. These marine wheels, enormous and accompa-nied by "swishing sounds," have been sighted in the East Indian archipelago and "bear an intriguing resemblance to the sounds reported during low-level auroras," according to Corliss, who adds that the luminous marine waves "emulate the auroral fogs." No one knows why these radiant mists should occur mainly in the Indian Ocean. But given a major seismic chain running under those waters (and hundreds of volcanoes in Southeast Asia), is it any wonder that the hidden forces of the earth erupt there? (This is the same area of the disastrous tsunamis of 2004.)

Dame Science remains silent on the subject of that awesome power. Nonetheless, the vortexyan powerhouse that envelops the earth, and the polar outcropping that is its nightside, is "very close indeed to our atomic energy." The extraordinary heat of the interior of volcanic fountains—over 2,000 degrees F—bespeaks atomic energy, as does the piezoelectric effect of earthquakes. The incandescence and rocket speed of ghost lights and mysterious fireballs also suggest a rarefied power. Indeed, the blue glow so frequently associated with earthlights may be traced to the chemistry of atomic nitrogen, as are the blue auroras. And if auroral wavelengths correspond to oxygen in its atomic form, so does night airglow—that soft, faint light that, absent moonlight, is brighter than all the starlight together. Does airglow come from the upper atmosphere, as science teaches? Perhaps it is the planet's gentle and diffuse out breath, quietly oozing out the earth from dusk to dawn. A "fictional" view of inner earth, envisioned by the prophetic Jules Verne in his *Journey to the Center of the Earth,* pictures our planet's interior "lit up like day . . . the illuminating power . . . flickering [and] evidently electric; something in the nature of the aurora borealis . . . light[ing] up the whole of the ocean cavern." His ideas, far in advance of the times, were one part imagination and one part ear to the ground. When a thirteenth-century Norseman speculated that auroras "radiate at night the light which has [been] absorbed by day"; when Scandinavian folklore insists that auroras originate from the depths of the sea; when Eskimos tell of auroras so low they have killed people—shall we dismiss their folk science as superstition?

Though many have dreamed of drawing power (free energy) from the earth's magnetic envelope, precious few have warmed to the secret of the northern lights. And with auroras romanticized as "a breath from beyond this planet," commentary on the auroras flip-flops from sentimental doggerel to scientific dogma. In fact, the "solar wind" theory bogs down when we take a look at the Van Allen Belt, where the northern lights are supposedly manufactured by solar radiation. But—have a look—the belt girdles the earth like a donut, straddling every latitude except the poles (where the earth's vent stream cuts its own swath into the atmosphere). If the northern lights were rained down to earth from the "Aurora Factory in the Sky" (Van Allen Belt), they would not cover the poles but instead would head for the equator and mid-latitudes. But they don't.

What is the sciencespeak that will emasculate this obvious fact? The chimerical solar wind is supposedly empowered by something called "Alfven waves," which "accelerate particles down from space," nicely accounting for the fact that auroral rays are actually "more intense as [they] converge near earth"! And "fade away rather gradually" into the atmosphere. But NASA has it the other way around!—filmed at Antarctica, its footage shows "the aurora emanating from an orifice in the continent . . . spouting out from the opening [and] shooting up luminous streamers . . . directed toward the zenith." Indeed, physicists, led by the giant Karl Gauss, recognize the magnetic nature of the earth's interior, arguing that the earth's force field is in fact generated by electric currents in the interior.

So why haven't we connected the dots?

Madame Science, for her own reasons, prefers to remain mum on the simple question of why auroras are nocturnal. Why can't they, like rainbows, occur at any time? Do Alfven waves/solar wind sleep by day? Or—was Ray Palmer right, after all, in calling our attention to that nocturnal energy which "flows upward from the north pole"?

Have you ever wondered why most earthquakes and volcanic eruptions occur at night or in early morning?

Now you know.

What about the twenty-four-hour delay between "solar flares" (read: vortexyan surges) and exceptionally large auroral displays?

Now you know, for that same energy must first pass through the earth in the nightwatches before streaming out the polar vents. And it has nothing to do with the sun.

The natural earth receives her energy by day and discharges her powerful magnetic flux by night. Is this "anomalous" (and therefore beneath the purview of science)? Shall we banish these lights, sounds, motions, smells, flames, surges, and commotions of our dynamic planet to the Siberia of scholarship, to do time with other rogues and mavericks of "fringe" science? In place of a thousand and one "sophisticated" explanations of Dame Science, a few, very few, principles will surely underlie the great panorama of earth's mysteries.

Geysers are tapped to heat the Icelandic city of Reykjavik.

Volcanoes, which release the energy of atom bombs, have proved an immense source of cheap power.

Huge (natural) steam bath galleries heat the towns of Antarctica with hot underground water.

And the auroras, one of the wonders of the world—the great light show in the sky—may yet prove a greater wonder when their awesome power is tapped for the good of humankind.

15 Techno Invisibility

Can Newly Emerging Technology Make
Solid Objects Vanish?

John Kettler

While natural camouflage, from two French words meaning "to cover with flowers," has been with us in nature for eons in creatures as diverse as the chameleon, the octopus, the zebra, and the flounder, organized camouflage in warfare, as opposed to tribal-type activities, didn't really get going until much later.

The famous Japanese Ninja, for example, date back to 600 CE, though the core ideas go back thousands of years to a Chinese book (not the man) called *Sun Tzu*. It wasn't until after the Boer Wars at the turn of the 1900s that the premier army of the world, the British army, ceased going into battle clad in bright scarlet uniforms and adopted the same khaki that had given the Boers such an edge in Africa, this despite an earlier savaging during the American Revolution at the hands of rifle-armed American snipers wearing hard-to-see buckskin.

The response was typically British: measured, controlled, understated, being expressed as a few specialist skirmishing units, notably the 95th Regiment, the Rifles, perhaps familiar to some readers via the "Sharpe's Rifles" PBS series, starring Sean Bean as an NCO and later an officer in the famous "Green Jackets," who first saw action in Spain during the Napoleonic Wars. For European fighting, green was a huge improvement over scarlet.

Although disguising ships as innocent vessels or even the other side's own was used for centuries, it wasn't until the American Civil War that camouflage for ships emerged. And even then, it was a kind of haze gray scheme intended to hide blockade runners slipping amid the coastal fog and mist from the vigilant eyes of the blockaders.

But it wasn't until World War I that warships began to sport official camouflage paint patterns, intended to either mask the presence of the vessel or fool the enemy gunner by distorting his perception of what he was seeing, taking a page from the zebra's bold natural camouflage. And as aerial reconnaissance and, later, aerial attack became more and more of a problem,

Fig. 15.1. An attempt at optical stealth was made by the Germans in World War I, as demonstrated by their Fokker E III see-through monoplane.

to both those on the ground and above it, camouflage measures became common for both ground and many air units. Still, some fighter units, such as Jasta 27, Richthofen's Flying Circus, took to the sky in bright, lurid paint schemes intended to be seen for miles and strike fear in the foe, such as the brilliant red Fokker triplane flown by the "Red Baron" himself.

One of the truly novel ideas to emerge from World War I was the notion of "optical stealth," little known but first achieved by the Germans in 1916 by taking a standard Fokker E III monoplane and, instead of covering the

Fig. 15.2. The transparent cellon skin of the Fokker E III see-through monoplane made the plane difficult to see.

wooden frame with fabric and doping it, first painting the frame white. And all solid items, such as the engine cowling, fuel tank, weapons, and so on, were either painted white or given a mirror treatment before the transparent cellon (cellulose butyrate acetate) skin, the brainchild of the little-known German Jewish chemist Arthur Eichengruen, was applied.

This unsung stealth pioneer is beginning to emerge from the mists of history, thanks to an essay in *Angewandte Chemie International** by Elisabeth Vaupel of the Deutsches Museum, Munich, FRG. Later, in the 1930s, Russian designer S. G. Kozlov (www.aviation.ru/okb.php) did basically the same thing with a Yakovlev AIR-4 parasol-winged trainer, but covered his with rodoid, a French-made transparent thick cellophane, which is why it was described as "organic glass." At medium and long range, the optical stealth worked great, but the problem was that airplanes were shifting to metal construction, with highly stressed metal skins, and were becoming more and more solid to boot, so the project was canceled in 1935.

Remember the Russians, though, for from them will come a huge stealth breakthrough. World War II was a time of great innovation in the field of techno invisibility, with certain aspects of it still being hotly debated. The knowns are astounding in their own right, including the German development of the all-but-invisible-to-radar Horten flying wing aircraft (Ho IX V3) that nearly made it into production, and the deployment of a primitive form of radar-absorbing material (RAM) on the snorkels of late-model U-boats. Add to this the American work on the "Yehudi effect," in which lights placed along the leading edges of the wings and engine cowlings of antisubmarine aircraft were used to remove the contrast between the dark aircraft and the bright sky, which was one of the key means of visual aircraft detection, the object being to prevent visual detection until the plane was two miles out, so close the U-boat couldn't submerge in time to escape.

While most don't know about many of these developments, where things get controversial in a hurry is when we look at the "other" American techno invisibility project, the brainchild of the great Nikola Tesla and the Princeton brain trust known as the Institute for Advanced Studies, consisting of people like Einstein and von Neumann, to name but two. This was the ship invisibility program commonly known as the "Philadelphia Experiment" but properly known as RAINBOW. Devised in response to so many ship sinkings by U-boats that at times they outpaced the ship-building rate, the idea was, through strong rotating magnetic and microwave fields, to

---

*Elisabeth Vaupel, *Angewandte Chemie International* 44, no. 22: 3344–55.

make a ship optically invisible by creating a kind of energy bottle around the vessel through which light could not pass. Some think this was also designed to defeat radar as well.

Allegedly, there were many wholly unexpected effects, such as teleportation. There were other negative effects, some expected by Tesla, which he gave warnings about, before quitting the project (reportedly sabotaging it out of concern for the crew) after his concerns were ignored in a headlong rush to demonstrate the concept on a crewed vessel. These had to do with sophisticated concepts of how humans fit into local reality and may be thought of, in *Star Trek* terms, as beaming partially into a solid object. Not good!

Though this may sound a bit far-fetched, some evidence suggests that not only did von Neumann carry on the work, but that the capability was eventually weaponized as well. We know, that in the 1980s, a U.S. aircraft carrier, dogged at every turn by Soviet spy trawlers, simply vanished in an instant only to reappear hundreds of miles away. Intelligence sources related the Soviet description of what happened when they lost track of a key American mobile nuclear strike platform as being "like someone stomped an anthill." Reportedly, everything that could fly or sail was pressed into a panicked, all-out hunt for the missing carrier, which was found way out of its original area only after many grueling hours of search. It is said that the American skipper got reamed for activating some highly classified technology without permission.

And while the Americans earlier learned the hard way in the skies over North Vietnam how deadly modern air defenses could be, a lesson relearned by the Israeli air force in the 1973 Yom Kippur War, few realized that a major key to the puzzle had been published in the Soviet Union in 1962, buried deeply in a body of work so poorly understood that the scientist himself said of his colleagues, "They thought my work was crap." And what was this marvel?

"Metod Kraevykh Voln v Fizicheskoi Teorii Difraktsii" ("Method of Edge Waves in the Physical Theory of Diffraction," *Moscow: Sovetskoe Radio,* 1962) by Pyotr Ufimstev. It is now available as DOC ID 19720010515N(72N18165) *NASA Technical Reports,* Report #AD-733203 FTD-HC-23-259-71 (243 pages).

It was directly from this deep technical paper that in 1975 Dennis Overholser, of what was then the Lockheed Skunk Works, gave his boss, Skunk Works president Ben Rich, the key to radar stealth in the form of an unprecedented plane designed as a series of carefully calculated plates, each forming a facet in a revolutionary design. It was so not expected to fly that it was dubbed the "Hopeless Diamond," but its real name was Have Blue, and it was the technology demonstrator aircraft for what became—after long, expensive, deep black development and a somewhat inglorious start in Panama during Operation Just Cause in 1989—the F-117 Stealth Fighter, the arguable star of Desert Storm in 1990. And it was in 1990, during a scientific visit to the United States, that Pyotr Ufimstev learned that his discoveries—which were ignored in his own country at the time (the Soviet Union collapsed in 1989)—had been eagerly and secretly seized upon by the Americans to field one of the key technologies generally credited with causing the Soviets to spend themselves into oblivion in response. How splendidly ironic!

It is against this background and such later projects as the B-2 Spirit Stealth Bomber, the Sea Shadow Stealth Ship, covertly built, interestingly enough, inside the notorious Glomar Explorer, and the fascinating reports of the Alien Replica Vehicle (ARV) whizzing around Area 51* that we need to look at two new developments, neither of which requires antigravity technology.

METAMATERIALS, NOT MAGIC

Some readers may recall a popularly reported "invisibility cloak" invented in Japan by Susumi Tachi. It worked on the principle of "reading" the scene behind the wearer, then projecting it onto the front of the cloak, thus hiding the wearer to a substantial degree via shape obscuration and through contrast reduction with the optical background. The invisibility cloak suffers from its sheer complexity, what with all those cameras, wires, light emitters, and much more. But suppose you could get the same effect—sans all the expensive electronics? What if the entire process could somehow be made entirely passive? It seems that what the Japanese inventor was working on, at a macro level, may have been a full-blown technical breakthrough at the micro and even nano levels. How Hermetic can it get?

The term *metamaterial* was invented by Roger Walser, of the University of Texas, and was used in a paper he wrote that was published in 2001. He defined it as "artificial composites . . . that achieve material performance beyond the limits of conventional composites." Unsurprisingly, this turned out to be of such interest to DARPA (Defense Advanced

---

Project Redlight DVDs at www.hourofthetime.shoppingcartsplus.com/page/page/2569223.htm.

Research Projects Agency) that it launched its own program in 2001 and prompted the agency, in the form of now Program Manager Valerie Brown and Stu Wolf, to expand the definition, as follows: "Metamaterials are a new class of ordered composites that exhibit exceptional properties not readily observed in nature. These properties arise from qualitatively new response functions that are: (1) not observed in the constituent materials and (2) result from the inclusion of artificially fabricated, extrinsic, low dimensional inhomogeneities."

Putting this into plain English, metamaterials are created by taking familiar composite materials and modifying them by adding microstructures or even nanostructures, yielding materials with properties so "out there" as to fall under Sir Arthur C. Clarke's dictum "Any sufficiently advanced technology is indistinguishable from magic." How "out there" are we talking? Think of it this way. If standard science were to be assigned only one quadrant of what's theoretically possible, metamaterials give us the other three quadrants. How's that for progress? This concept is explored in David R. Smith's "Electromagnetic Metamaterials" portion of his site.* Smith is one of three authors of a key paper on metamaterials published May 25, 2006, in *Science*.

The real excitement lies not in sharpening antenna patterns, making better motors, and what not, but in the stunning stealth possibilities made possible by metamaterials—specifically, bending microwaves and light to cloak something that needs to hide, be it an aerospace craft or a GI. This is what the *Science* paper was all about, and Smith really goes into the possibilities in a piece called "Blueprint for Invisibility."† There, he looks at the fundamental notion of invisibility, as expressed in science-fiction books and films, then compares and contrasts the shortcuts allowed writers and film-makers with what's doable, and potentially doable, in hard science. It can't be done yet, but here's where the technology is going.

WHEN IT'S OKAY BEING WARPED

What if you could arrange it so a radar or visual observer saw not the aerospace vehicle/individual armored vehicle or warrior but what was behind it/him/her, as though the actual target didn't exist at all? Wavelength issues mean that this technology will first be seen in the microwave region, for the

---

*At www.ee.duke.edu/~drsmith/about_metamaterials.com.
†At www.ee.duke.edu/~drsmith/cloaking. html.

wavelengths involved are much shorter than those for visible light, but it, too, is at least theoretically possible. So, how would it work?

As noted earlier, metamaterials work in ways we've never seen before, which include exhibiting the remarkable and previously unseen negative index of refraction when struck by electromagnetic waves of the appropriate frequency. In simple terms, instead of refracting the energy toward the source, the material would turn the energy away from it, much in the manner that water flows around a pier or bridge abutment. The concepts reported to date do this with either tiny spheres or cylinders carefully built right into the composite at either a microscopic or a nano scale. They basically catch, then flow, the incoming microwave energy or light, depending on the design, around the viewed side and across to the opposite side, where it exits.

From the observer's viewpoint, the item simply isn't there, for there's nothing to perceive, at least within the designed coverage of the metamaterial. If it's a radar, the signal simply keeps going until it hits something that is radar reflective, then returns and is measured in the usual way. If it's visual, then the observer or vision system sees only what's behind what would otherwise be a target.

Is this a panacea? Hardly. The metamaterials are "lossy"—that is, they tend to absorb upward of 20 percent of the original input signal; they are also highly frequency specific, meaning that for each frequency band of interest, a different variety of tailored metamaterial will be needed. There is presently no known broadband metamaterial.

Interestingly, there's a parallel between metamaterials and modern electronics. Both rely on the deliberate contamination of an otherwise innocuous, everyday substance to do things otherwise impossible.

16 Weather Wars

Is There an Unnatural Side to Natural Disasters?

John Kettler

The historic 2005 tropical storm season (in which there were so many storms that storm specialists ran out of the allotted list of names and had to use Greek letters instead) appears to have finally jarred the mass consciousness into realizing there's something fundamentally off with the weather. For "off," substitute "unnatural." Consequently, concepts and people normally kept on the sidelines and out of the public eye are receiving unprecedented coverage and exposure in such unlikely places as *Business Week Online* magazine.

Consider the following: First, "normal" weather is, from a long-term-climatological perspective, an exception to the general pattern; historically it has been both generally colder and far less stable. It seems we've been living the halcyon days and have gotten used to them.

Second, the global warming theory based on the greenhouse effect is merely a theory. No amount of scary publicity or punditry will make it otherwise—glacier recession notwithstanding. Many reputable scientists do not subscribe to the theory at all and have published withering critiques of same.

Third, many have argued that it's not the atmosphere that's heating up, but instead the planet's interior, evidenced by eruptions from long-dormant volcanoes, the appearance of new volcanoes, and numerous reports of building volcanic forces from earth sensitives.*

There are several other players, too. The sun has been highly and abnormally active outside of its eleven-year cycle, and Mitch Batros, at Earth Changes TV (www.ectv.com), has established a direct link between solar upheavals and crazy weather, among other things. Evidence shows that stellar instabilities ("starquakes") directly affect this planet, and we haven't even discussed the planet X/Nibiru scenario and the potentially

---

*See "The Bio-Sensitive Factor," *Atlantis Rising* 27: www.earthboppin.net/talkshop/feelers.

huge weather effects as it, allegedly, nears Earth. Some even believe we could well have global hyper-weather like that portrayed in the upcoming TV disaster film *Category 7: The End of the World* or in the movie *The Day After Tomorrow*.

And with all of the above going on or lurking about, could we also have people messing with Mother Nature?

CONNECTING WITH MOTHER EARTH

Commentators ancient and modern have long noted an apparent relationship between fierce battle and extreme weather, with the scientific explanation being that all the noise, dust, and so on act to trigger greater than normal precipitation via enhanced condensation effects and, perhaps, adding energy to a nascent storm. This purely mechanistic explanation, of course, leaves Mother Earth, Gaia, out of it. Is this wise? To those schooled in the Hermetic tradition ("As above, so below"), the answer is no!

Interestingly, in regard to Hurricane Katrina, we find a remarkable congruence between what some call the "fundagelicals" (a coined word for fundamentalist evangelicals) and their most unlikely brethren, the Luciferians. The common theme? Punishment. The fundagelicals view the destruction of New Orleans as nothing less than divine retribution against the city for its institutionalized wickedness, and the Luciferians, in the person of psychic Aaron Donahue, view it as earth's ". . . rising up and bringing judgment after years of misuse and abuse by mankind. As the planet is dying, it (she) is using her energies to attack the people who continue to kill her without thinking." As reported in "Aaron: Katrina—A Mother's Gift to a Greedy Nation" (www.farshores.org/jd090305.htm), he told Luciferians listening to his *Voice of Lucifer* Internet radio show to "learn to live close to the earth to be protected from the chaos that is befalling mankind, not just in the United States, but all over the world."

Sounds weird, but it's simple to show the same theme has been offered time and time again, not only in cultures girdling the globe but also in dreams, visions, and prophecies as well, given to people as religiously and ethnically diverse as tribal chieftains and suburban housewives. The warning is clear and unambiguous: Humanity's vibes and behaviors are faithfully reflected by Mother Earth. The crazier and more destructively we behave, toward the planet and each other, the more unstable and dangerous our loving host becomes. But what if people are deliberately messing with the weather? Is that even possible?

OF POLITICS AND SUPERWEAPONS

Secretary of Defense William Cohen, at an April 1997 counterterrorism conference sponsored by former senator Sam Nunn, said, "Others [terrorists] are engaging even in an eco-type of terrorism whereby they can alter the climate, set off earthquakes, volcanoes remotely through the use of electromagnetic waves . . . So there are plenty of ingenious minds out there that are at work finding ways in which they can wreak terror upon other nations . . . It's real, and that's the reason why we have to intensify our [counterterrorism] efforts."

As noted by retired Army lieutenant colonel, towering intellect, energy researcher, and scalar weapon "prophet" Tom Bearden on his in-depth site (www.cheniere.org), the bracketed material in Secretary Cohen's quote is textual "clarification" (read: spin) by Pentagon PR types. The original statement means groups and nations alike have this devastating electromagnetic weaponry at their disposal.

And what groups!

Independent and also collaborative research by Tom Bearden and veteran petroleum geologist Ron Mason has uncovered an alliance unholy even by Hollywood-villain standards. In a six-part series in *Nexus* magazine (www.nexusmagazine.com) titled "Bright Skies,"* Mason makes a cogent case that the Aum Shinrikyo "Divine Truth" sect (of Tokyo subway "sarin" attack notoriety) quietly built and tested in the Western Australian outback a Tesla-based device that not only triggered quakes but also caused them in an area devoid of them in both historical records and aboriginal tribal memory. Later articles in the series amended this view to indicate that the scalar shots were fired from Russia into Australia. Worse, it was said, the firing trials were on a grid pattern, with no regard whatever to underlying geologic structures. Ah, but that's not the really bad news. The ultranationalist sect has powerful allies, you see.

Terrorist groups require funding, so it seems only natural that an ultranationalist group aching for vengeance against the United States for the Hiroshima and Nagasaki atomic strikes would get what it needs from a like-minded entity. How about the dread Yakuza (Japanese Mob), whose origin was reportedly a group of diehard ultranationalists at the end of World War II? Now, there's a group with money and power. But superweapons, where to get superweapons? Quick. Name a former superpower that collapsed in 1989. Now, name its once all-encompassing security organ, an organ that

---

*See articles, video, and links: www.cheniere.org/misc/brightskies.htm.

controlled not just nuclear warheads but scalar weapons, too. Did you guess Russia and the KGB (now FSB)?

According to the research by Tom Bearden, Ron Mason, and others, that's the core scenario we're facing. Prompted by both hatred for the United States and a large down payment ($900 million in gold), KGB/FSB members have quietly leased earlier-generation Russian scalar weapons to the Aum Shinrikyo/Yakuza axis, and that entity is, among other nasty, terrifying things,* waging weather war against the United States, and getting rich in the process.

How? By investing in energy futures, then torpedoing the market with huge preplanned dislocations (targeted hurricanes that wreck offshore rigs and shut down refineries), causing prices to soar. The closure of the Port of New Orleans for a time deprived the United States of 20 percent of its port capacity and hamstrung critical imports and exports. As a direct result of Katrina's devastation, heating oil prices rose, hitting the pockets of millions of Americans. And these are reportedly but the tip of the cascading economic havoc. Nor is this the first such strike. In part five of "Bright Skies," Mason cites a string of weather attacks directed squarely at Western countries and their allies—old news to Tom Bearden, who dates such attacks clear back to the Cold War and the operational weather modification use of the Russian "Woodpecker" (combined over-the-horizon radar system and scalar weapon): "North America has not had 'normal' weather since 1976." Perversely, the Russians chose July 4 to initiate Woodpecker-based scalar weather modification.

Another disturbing angle to this scenario is that Mason turned up evidence that the Aum Shinrikyo deputy may have been involved in a covert Japanese scalar weapon program in Kobe, Japan, a program literally demolished by the January 17, 1995, Kobe 7.2 quake, following a "prophecy" from the head of the sect. It further appears that the Yakuza breached KGB/FSB security on the leased weapons and learned the technology—to the point that it is now making its own portable scalar weapons.

Admittedly, this is "out there" by most people's standards, but as noted earlier, it's drawing mainstream attention. The coverage tone is both dismissive and derisive, but that the topic and some of the key players are being covered at all is amazing in and of itself.

---

*For the chilling details see "The Quantum Menace," *Atlantis Rising* 50, Bearden's *Fer De Lance*, 2nd edition, and the earlier weapon slides at www.cheniere.org/images/weapons/index.html.

In a licensed *Business Week Online* story titled "Who Controls The Weather?" (www.wtov9.com/money/5141496/detail.html) Tom Bearden and the basics of the scenario are fairly accurately reported in the first two paragraphs, some supporting sites are listed in the third (with the term "conspiracy theories" bandied about), followed by a searing dismissal in the first sentence of the fourth paragraph: "To almost all scientists and weather professionals, this sort of rationalization is ludicrous." Good thing that writer used a qualifier, for at least one former professional TV meteorologist hasn't gotten the word, and he, too, is causing a buzz.

Fig. 16.1. Hurricane Katrina bears down on New Orleans in 2005. (Satellite photograph courtesy of N.O.A.A.)

Pocatello, Idaho, is not generally considered to be a hotbed of controversy, but Scott Stevens and his site Weather Wars (www.weatherwars. info) have made it so. Drawing on not just Bearden's work, but also information, videos, and photos he and others have gleaned, he presents, in a straightforward manner, the astounding evidence for the unnatural character of not just Katrina, but many other "impossible" and "unprecedented" hurricanes as well. A *MosNews.com* (as in Moscow, Russia) article, "U.S. Meteorologist Says Russian Inventors Caused Hurricane Katrina," cites a *Village Voice* interview with Stevens "right after Katrina's landfall," which was picked up by *Wireless Flash*. His assessment was unflinching: "There is absolutely zero chance that this is natural, zero." He feels so passionately about what he's doing that he recently resigned from the TV weather job he's held for years in order to devote full time to his research and exposing the truth. Naturally, he's getting all kinds of flak, being called "nut job" on certain Internet sites and "delusional" by a columnist for the *Chicago Tribune*.

"WEATHER MADE TO ORDER"

In 1992 the *Wall Street Journal* reported that a Russian firm bearing the fascinating name Elate Intelligence Technologies Inc. was offering weather modification for hire, with offices near Moscow's Bykovo Airport, so close that the firm's special antennas could be seen from there. Various sources indicate that such weather modification could be done over "200 square miles." The firm's director, Igor Pirogoff, whose quote forms the subhead above, was in no doubt as to what his company could do, saying it could've turned Hurricane Andrew ($17 billion in damage) into "a wimpy little squall." This individual also pioneered a new concept in perversity, "weather extortion": "We guarantee good weather for your outdoor event if you hire us—and bad weather if you don't pay as agreed."

As beautifully set forth in "Stormy Weather: The government's top-secret efforts to control Mother Nature,"* writers Bob Fitrakis and Fritz Chess show that the United States not only has understood, since the fifties, the range of military advantages associated with weather modification and control, but also has poured considerable resources, for many decades, into developing and fielding an array of operational capabilities. One such early use, Project Popeye, was seeding clouds with silver iodide over the Ho Chi

---

*See www.rense.com/general18/mn.htm.

Minh trail with the goal of washing it out, while destroying crops used to support the vast logistical network it encompassed. Reportedly, rainfall went up 30 percent. Unexpected precipitation came later in the form of outraged Senate hearings and a UN treaty.

On December 10, 1976, only a few years after Project Popeye, the alarmed diplomats composing the General Assembly of the United Nations approved the *Convention on the Prohibition of Military or Any Other Hostile Use of Environmental Techniques.* Article I.1 requires each State Party to the convention "not to engage in military or any other hostile use of environmental modification techniques having widespread, long-lasting, or severe effects as the means of destruction, damage or injury to any other State Party." Article I.2 commits the signers to not encourage States, groups thereof, or international organizations to do so either. Article II defines environmental modification as "any technique for changing—through the deliberate manipulation of natural processes—the dynamics, composition or structure of the earth, including its biota, lithosphere, hydrosphere, and atmosphere, or of outer space." Article III specifically exempts peaceful uses in accordance with established principles and international law. Okay so far, but what about, say, individuals, companies, and organizations? Article IV requires the signers to control illegal environmental modification activities "anywhere under their jurisdiction or control." Article V mandates consultation and cooperation but has no teeth.

Prior to treaty signing, the then Soviet Union had commenced weather war, a gross treaty violation, against the United States, which meanwhile, it seems, was quietly pursuing all sorts of classified weather modification projects of its own.

HAARP

The U.S. High-Frequency Active Auroral Research Project (HAARP), primary facility, in Gakona, Alaska, supposedly grew out of a desire to use the vast Alaskan natural gas resources to generate power and beam it to users, rather than build a pipeline, with all that entailed. Its DoD cover activity is upper ionospheric region research, but—as shown by researcher Bob Fletcher in his special report "Weather Control as a Weapon" and Dr. Nick Begich and Jeane Manning in their seminal book, *Angels Don't Play This HAARP*—the reality is appalling. From an environmental modification standpoint, HAARP can: a) move or block the jet stream, changing weather over entire regions, b) artificially load fault zones, triggering earthquakes,

c) energetically alter space, and d) use ionospheric lenses and mirrors, aided by pervading the sky with things like toxic barium compounds via chemtrailing, to not only interrupt or knock out all communications, power, and electronics in a selected area, but also to disrupt the nervous systems of people and/or other living things, killing them at will if desired. Alarmingly versatile, HAARP can affect every single area and item prohibited by the treaty.

So next time a "natural disaster" strikes, must we now ask: "Natural, unnatural, or artificially assisted?" And must we also ask: "Who benefits?"

PART FOUR

SPIRITUAL SCIENCE

17 The Sensitivities of Water

Startling New Evidence That Water Can Reflect Thought and More

Jeane Manning

Does water record the themes of our thoughts and feelings? Do water crystals dance to Mozart and fall apart when subjected to heavy metal lyrics? Can water reflect the power of unconditional love and mirror the effects of gratitude?

A researcher in Japan believes he has photographic proof of water possessing such sensitivities. Dr. Masaru Emoto studies the microscopic architecture of tens of thousands of water crystals. When healthy water freezes, it creates crystals—solids with an orderly internal structure. But the ability to create crystals can get knocked out of water. Emoto sees those structures as influenced by human activities and intentions. He sees physical proof that our thoughts—human vibrational energies—affect our surroundings. Further, he sees that music and even pictures affect the molecular structure of water.

Emoto visited British Columbia recently, told his story, and presented new information he's learned after the publication of his book *Messages from Water*, of dozens of extraordinary photographs. He had received a doctorate in alternative medicine from the International Open University in Sri Lanka in 1992. His background is as a healer who treated more than 15,000 people and as an author of a dozen books about subtle energies.

Emoto first turned toward researching water after meeting with Dr. Lee H. Lorenzen. Lorenzen, who had studied biochemistry at UC Berkeley, developed a type of water called "Magnetic Resonance Water" to heal his own wife's health problems. Through Lorenzen, Emoto got his hands on a machine called a Magnetic Resonance Analyzer (MRA) that was said to measure *chi*. When speaking of chi, Emoto uses the Japanese word *hado*, meaning "the world of subtle energy related to consciousness."

Back in Japan, Emoto says, he learned that "hado water" produced by Lorenzen at Emoto's request improves people's health. His patients' families could see the improvements, but skeptics scoffed at the suggestion that water could hold and impart health-enhancing information. Who has seen chi? What does hado look like?

To show the skeptics that what was happening in the hado water was more than merely imagined effects, Emoto wanted a tool or method for measuring differences between one type of water and another. In 1994 he read a book that started him thinking about how he might be able to find such a method. The book noted that even after a few million years of snow falling on earth, as far as scientists can find there aren't any two snowflakes that are identical. This led Emoto to ask if freezing water samples would be a way to see information held by waters subjected to different influences.

He knew that homeopathic dilutions continuously hold information—about molecules of substances that had been added to water before the molecules were diluted out of the water. However, it would have been difficult for him to study homeopathically imprinted waters because homeopathy is not officially allowed in Japanese medicine. He decided instead to begin his experiments with pure waters.

Freezing droplets and photographing the resultant ice crystals individually turned out to be easier said than done. They had to be seen under a powerful microscope and photographed at high speed before they could melt. Nevertheless he persevered in developing a technique for clearly photographing crystals at magnifications of between 200 and 500 times in the ninety seconds before they began to melt. The technique included building what was in effect a large icebox in which his staff worked for up to fifteen minutes at a time while dressing in parkas for minus 5°C (23°F) temperatures.

Emoto laughs at himself while admitting he can't take more than a few minutes of such temperatures and therefore had to hire out the actual photography. He then enlarged the photos into slides and showed them to

Fig. 17.1. Renowned author and healer Dr. Masaru Emoto.

his students, who were fascinated at what they saw. For instance, one sample of snow-fed springwater from Yamanashi Prefecture gave a symmetrical hexagonal (six-sided) crystal with three branches stretching out from each edge, giving the impression of people holding hands together.

In contrast, exposure to chlorine has a shattering effect on water crystals. Emoto's students soberly concluded that since life on earth depends on water, the life force–related quality of water in any given place must have a tremendous effect on the environment.

Emoto's associates sent him samples of water from various parts of the world—polluted rivers and holy sites, various cities and mountains. His staff took many photographs of each sample. Although individual crystals from one water sample differed incrementally, crystals from any given sample were similar. Unpolluted water samples yielded symmetrical hexagons, but when polluted water droplets were frozen, the photographs revealed an inability for crystals to reach complete hexagonal structures.

However, most samples went through a common stage when the ice droplets melted. Just before ice turns to water, a shape could be seen under the microscope, which is a replica of the six lines inside a circle constituting the Chinese alphabet's character for water.

What caused the apparent weakness in the waters' hidden structures? The problems correlated with whether a water had been exposed to chemicals, emotional pollution (panic pervading a city after an earthquake destroyed the water's crystallizing abilities), and sound pollution.

In later experiments, Emoto put the water samples between loudspeakers and exposed them to certain recordings before freezing the droplets. When music lyrics contained aggression such as "I hate you!" or "You fool!" the water not only could not form proper crystals; under the microscope, it had a chaotic appearance.

On the other hand, uplifting music such as Mozart's Symphony No. 40 in G Minor and Beethoven's *Pastorale* yielded beautiful gracefully formed crystals. This line of research led to placing samples of water on certain photographs, attaching labels with words written on them, and having schoolchildren verbally give messages to the water samples.

Although healthy waters form a myriad of variations on hexagonal crystals, Emoto was in for a surprise. One of his before-and-after sets of photos showed inadequately formed water crystals from a sample of Fujiwara Dam waters before a minister prayed over the dam. Waters behind a dam become stagnant, and the sample under the microscope looked like a suffering face. However, the "after" sample was taken after Reverend Kato Hoki, chief priest of Jyuhouin Temple, Omiya City, performed a prayer practice for an hour beside the dam. Among the exquisite hexagonal crystals obtained from the water after it was prayed over were two heptagons—crystals that were seven-sided. Interestingly, the reverend had prayed to the Seven Goddesses of Fortune.

Emoto's research indicates that troubled tap water that cannot crystallize properly can be transformed to make beautiful crystals through conscious thought focused on love. He learned that the most powerful combination of words are "love and gratitude."

His recent experiments indicate that electromagnetic pollution can be eased if the words coming over the cell phone or the television are harmonious, such as a telephone call between lovers or a TV program about nature. Political debates in general had a negative effect on the waters' abilities to crystallize.

If water carries messages about the intentions—whether loving or angry—of those who handle it, what does that fact mean in our everyday lives? Until a variety of scientific studies are done, we can only speculate.

We've heard tales of house plants dying when their home became a marital battleground, and conversely plants thriving under the green thumb of a gardener who loves growing them. Perhaps water inside the plants instantly registers and records the powerful emotions emanated in their environment.

Were the mysterious giant cabbages grown in the early days of the Findhorn community in an inhospitable corner of Scotland partly a result of influences upon the water fed to them? A member of the Findhorn Foundation who recently visited British Columbia said the abundant growth of that garden had been a demonstration of what happens when

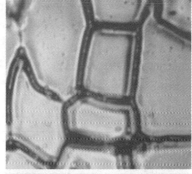

Fig. 17.2. The photograph on top is of stagnant dam water. The photograph on the bottom is dam water after it had been exposed to prayer. (Photographs from Messages from Water*)*

humankind cooperates with another realm (nature spirits), but perhaps water was part of the mechanism for manifesting the demonstration. Water could have been transmitting the purest intentions of the humans involved. The experiments of scientists such as Stanford University professor emeritus William Tiller, Ph.D., indicate that our intentions have measurable effects on physical objects.

Is water the most sensitive medium for recording and transmitting subtle influences? If so, the implications for our health and the health of our environment are profound. The fertilized human egg is about 95 percent water and a mature human body is 70 percent water. We live on a planet whose surface is about 70 percent covered with water.

Emoto has presented his findings to audiences in Europe and Japan. In England he met with Rupert Sheldrake, Ph.D. Emoto says he was asked by Sheldrake to do experiments involving people in jail, to see how the water influenced by people who are criminals or are generally in a negative state of mind differs from influences of people in other sectors.

Like all pioneers, Emoto's work will surely run up against mental blocks. But since it is self-funded, at least it won't be shut down, as Dr. Jacques Benveniste's laboratory was when the French government pulled his funding after his research seemed to vindicate homeopathy.

What seems to be needed now is for rigorous double-blind scientific experiments, impeccably designed attempts, to either replicate or refute Emoto's findings. However, Emoto wonders if the Western scientific model is capable of the job. When dealing with something alive such as he believes water to be when in a state of health, there is no such thing as an exact replication. Like snowflakes and human faces, he says, water crystals are creations of a myriad of influences and there will be no two identical.

How do you replicate the scenes in a kaleidoscope? Water is apparently so sensitive that it is an ever-changing record of subtle influences.

18 The Power of Water

Are Its Secrets the Keys to Solving Today's Most Vexing Problems?

Jeane Manning

Our thinking apparatus runs on water. Our physical bodies are two thirds water, so obviously its qualities can heal or harm us. We now learn that water seems to remember and later convey "information." No wonder the most dynamic frontier in science today is water research. Or is it a re-search, I wondered, after encountering researchers who:

- show how neuroscience tends to confirm medieval concepts situating memory, imagination, and reason in water-filled cavities of the brain;
- experiment with transferring, from water to us, the life-force energy chi, also called prana down through the ages;
- study specially shaped water pipes used by the ancient Minoan culture in Crete;
- show how the emanations from healers' hands change water;
- measure physical qualities of "holy water," or effects of conscious intent upon water's crystalline structure; and
- build prototype inventions aimed at using water as a source of energy.

Some study the big picture, such as the claim that rivers self-organize and energetically recharge themselves through spinning motions. And some point out the well-known anomalies—water is densest at 4 degrees Celsius (39.2°F) and strangely expands when it cools further, so that its solid state floats on top of its liquid state. Water as the "universal solvent" melds with nearly any other element. Water's main ingredient, hydrogen, is spread throughout galaxies, and ice is found in dust clouds in outer space.

The picture of water that emerges is what *Aquarian Conspiracy* author Marilyn Ferguson calls "the strangest stuff around." Learning about the mysteries of water evokes a primal foreknowing, like a racial memory, Ferguson wrote a few years ago. "Prescience, perhaps pre-science, something we've known for a very long time."

Before our materialistic age lost the abilities to sense subtle energetics, water was central to sacred rituals and symbols. Baptism. The holy river. Spiritual visions of the ocean of love. Myths of the Flood or of creation. Drinking of sacred waters when visiting an oracle or a shrine. The Sumerian goddess Inanna had a vase in place of a heart from which flowed miraculous water. The Bronze Age civilization of King Minos at his city of Knossos on the island of Crete apparently lived by the principle that water should be returned to the earth in the same condition as it was when it was borrowed—treating all water as holy. Our era, in contrast, treats rivers and oceans as dumping grounds, and we face shortages of drinkable water. Dr. Karl Maret predicts that water will be the currency of the next century. Meanwhile, researchers of water's mysteries struggle for funding.

"The quest to understand water hasn't summoned up the capital and glamour of space research," Ferguson notes, "although it may have more direct bearing on our lives." While humans burn rain forests and alter other factors that kept our habitat moist, "we should remember the nagging suspicion that Mars was once a watery planet."

LET WATER MOVE; KEEP IT COOL

We've had ample warnings. In the twentieth century Austrian forest warden Viktor Schauberger (1885–1958) warned about wastelands appearing on our planet when vast forests disappear. He observed water's interaction with the forest, such as the vitality of cold, pure water in tree-sheltered streams. "Comprehend nature, then copy nature," he admonished. He taught that water is a living rhythmic substance. In maturity it gives of itself to everything needing life. However, water can become diseased through incorrect handling. Dying water harms animals, plants, and fish.

Whether stilled by a dam or a bottle, stagnant and warm waters begin to deteriorate. Conversely, at a cool 39 degrees F, moving water is densest, strongest, and at its best carrying capacity. Wild rivers have inherent self-control mechanisms if left alone to establish their own homeostasis—that is, if kept cool with natural overhanging vegetation and allowed to meander around bends and therefore be lively with purposeful swirling motion. Short-sighted human engineering—clear-cut forests, megaproject dams, and rivers confined into canals—tampers with the circulatory system of our planet.

Having interfered with the hydrological cycle, we reap floods, droughts, and other extremes of weather.

The book *Living Water,* by Olaf Alexandersson, introduces Schauberger's insights into river management, water-fueled devices, and energy. Its successor is the book *Living Energies,* by Callum Coats, which could be the textbook for a new eco-technology—how to construct or encourage processes that don't fight nature but instead work in harmony. Coats researched for two decades into Schauberger's discoveries—from forestry to flood control to soil fertility and water purification. Reading this book, hydrologists could learn how crucial are small variations in temperature in a river. Among Schauberger's observations were how water's spinning motion recharges it with subtle energies.

WATERPOWER WITHOUT DAMS

The naturalist's warning echoes across the decades, "Prevailing technology uses the wrong forms of motion." Twentieth-century machines leave behind waste products because their processes use the destructive half of nature's cycles of creation/destruction—the centrifugal outward-moving motions of heating, burning, pushing, radiating, or exploding. They channel air, water, and fuels into the type of motion that nature uses to decompose matter. Schauberger observed that the centripetal inward-spiraling force is the creative, cooling, sucking motion without friction that results in increased order instead of destruction. He applied his understanding of cycloid spiral motion to a wide range of inventions and methods that are in harmony with nature's creative motion. The "water magician" had

Fig. 18.1. Nature pioneer Victor Schauberger, whose understanding of the nature and dynamics of water still informs environmentalists today.

solutions for agriculture and energy generation, as well as for transport of water in pipes that encourage water's inward-spiraling motion.

Schauberger's knowledge is sparking experiments by today's researchers. For example, some Scandinavians called the "Malmö group" use the phrase "self-organizing flow" to describe what they are creating, since Schauberger's technology made use of the natural orderliness spontaneously created by a system under the correct conditions.

Meanwhile, new energy-generating processes such as Dr. Randell Mills's BlackLight Power convert ordinary water into hydrogen and oxygen. Paul Pantone of Utah runs engines on water mixed with waste substances, and the air that comes out of an exhaust pipe won't dirty a white handkerchief held on the end of the pipe.

About a century ago, John Worrell Keely figured out how to run a motor on the power of cavitation, or implosion, while alternately compressing and expanding water. He harnessed a phenomenon that we dismiss as a nuisance—the "water hammer" in water pipes. Dale Pond, researcher of Keely's physics, says that Keely's Hydro-Vacuo motor created a water hammer shock wave that, when synchronized with the wave's echo, "results in

Fig. 18.2. John Worrell Keely (1837–1898) of Philadelphia was a carpenter and mechanic who, in 1872, discovered a new principle for power production via his Hydro-Vacuo engine.

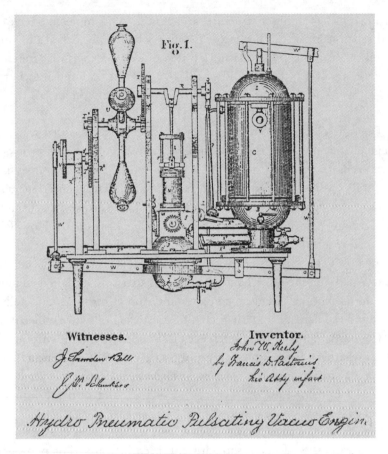

Fig. 1.

Witnesses.

J. Snowden Bell

E. M. Schumaker

Inventor.

John W. Keely
by Francis D. Pastorius
his Atty in fact

Hydro Pneumatic Pulsating Vacuo Engine

*Fig. 18.3. An illustration used as part of the patent for John Keely's
Hydro Pneumatic Pulsating Vacuo Engine.*

Amplitude Additive Synthesis—tremendously increased energy accumulations" in quick order. Pond warns that this resonance amplification is similar to the process that breaks wineglasses.

DO WE REALLY KNOW WATER? LIQUID MEMORY

At water-science conferences that this journalist attended in recent years—the November 1998 conference at Semiamhoo Resort in Washington state, funded by Living Water International; a privately funded 1997 meeting in Los Angeles organized by Linda McClain; and the Institute of Advanced Water Sciences (AWS) symposium the previous year in Dallas—one fact that emerged is that water is not a single homogenous product of nature. Water in living cells has a unique structure, and clusters of its molecules have organized relationships. Another factor is what Schauberger called the "immature taker" vs. "life-giving mature" water. Since water without minerals is a relentless solvent, if we could distill 100 percent of impurities out of a batch of water it would be dangerous to drink, leaching minerals from our bones.

Then there's the movement-vitality factor. Stagnant bottled water, even though chemically clear, is dead compared to water in rushing brooks. But water needs to move properly. As water is pushed through cities in the unnatural confines of metal pipes, its energetic oscillations interfere and the natural order in water's structure is canceled. How do we know this? For one, German engineer Theodore Schwenk and his Institute for Flow Science developed a technique for photographing the internal structure of water. In drops of water taken near pristine springs, a symmetric rosetta pattern was revealed. On the other hand, the internal structure of damaged municipal water is chaotic. Chemical contaminants and electromagnetic pollution compound the damage and cause chaotic clustering of water molecules.

The participants in the conferences I attended wrestled with questions such as whether "living water" is an organized state of matter and energy, capable of storing and transmitting information. If so, the implications go beyond homeopathy and "energy medicine" and into the interaction between water and consciousness.

Dr. David Schweitzer, grandson of Albert Schweitzer, is the first scientist to photograph the effects of thoughts captured in water! His photographs show that water can act as a liquid memory system capable of storing information. David Schweitzer first stepped on to this trail by becoming an authority on blood analysis. He learned that blood cells express themselves in sacred geometry and other harmonious shapes and colors. Since blood cells hang out in water, he looked further into that substance for answers about our thinking processes. After ten years of observing blood, in 1996 he made the discovery that opened the door to photographing the stored frequencies in homeopathics and natural remedies and to researching the impact of positive or negative thoughts on bodily fluids.

"Having studied the relationship between the brain, cells, and emotions," he told Joseph Duggan in Vancouver, "I came to realize that certain trace elements were needed to send information from one area of the brain to another." Minerals alone could not convey information. To find out if the carrier is water itself, Dr. Schweitzer experimented. French scientist Jacques Benveniste had already shed light on the memory of water in homeopathy.

He and a dozen other scientists demonstrated that water can retain a memory of molecules it once contained. In 1988 *Nature* magazine published their experiments showing that if water containing antibodies was diluted repeatedly until it no longer contained a single molecule of antibody, immune cells still responded to the water. The publication drew outrage from orthodox professors, and the magazine later sent a team to Benveniste's laboratory, including the magician James Randi and Walter Stewart, a self-appointed investigator of scientific fraud. The team judged the French scientists' results to be a "delusion." However, a recent book by Michel Schiff says the slander of Benveniste was the delusion.

Dr. Schweitzer says aspects of the homeopathic research couldn't be measured by the investigators' instruments. The witch hunt in France didn't stop him from radical thinking. He remembered Albert Einstein's idea that particulate "light bodies," also known as somatids, act in ways that we don't yet understand. Waking up one morning with insights on how to make these light bodies visible, Schweitzer began working on a fluorescent microscope at a certain intensity of light. He wanted to see somatids change in response to thought and other influences. Just before the water on the microscope slides evaporated, he saw certain formations develop "dependent on the thoughts or energy atmosphere it had been impregnated with. I observed that this cluster could be modified at will." Further work showed that microscopic light bodies in the water intensify in the presence of positive thoughts. They shine brightly if thoughts are backed up by emotion, and it makes a big difference whether the emotions are negative or positive.

Intrigued by the tiny light bodies, he tested holy waters of religious faiths from Italy, Russia, Yugoslavia, and North America and saw somatids floating even after years of being bottled on shelves. "This means there is an ideal balance with the somatids never touching each other, which gives them the greatest capacity to store information." However, in studying homeopathic remedies he learned that careful storage of energy medicines is crucial. French allergist Jacques Benveniste had learned that electronic circuits can impress lasting information upon water, and low-frequency electromagnetic radiation and heat destroy homeopathic strength. Further, Dr. Schweitzer has a warning about purified water that we buy in clear plastic bottles that have been exposed to fluorescent lighting. When we drink only this water, our lips dry out and become chapped and cracked. "Normally, drinking water does not dry out the mouth, but fluorescent lighting changes the structure of water such that it dries out the mucous membranes."

Randy Ziesenus, of Edmund, Oklahoma, says anyone can personally improve the water he or she uses. "It's amazing what happens when you just take a glass of water and hold it between the palms of your hands and ask your higher Self to work with that water and whatever you need for your highest good. And then drink it; incredible what that little [ritual] does." Ziesenus is president of Bio-Com, a company that specializes in development of biotechnology using radio frequencies (RF) to alter water's bonding structure. He says, "[I]f you drink water that's harmonious to the human body, the water will pass through the body within ten to fifteen minutes. Then you've got to go to the restroom. The [harmonious] water will carry out toxins."

One of his inventions condenses water from air. "That's one of the biggest things I've been working on—using frequencies to draw moisture out of air." He and researchers from Los Alamos National Laboratory are working on "a program where you can take a photocell device, put it out in the desert, and it'll make a gallon of water overnight." The unit is powered by photovoltaics (electricity from sunlight). Ziesenus agrees with Dr. Schweitzer's claim that our AC electricity leaves a harmful imprint on water.

WILLIAM TILLER

At the Living Water conference, Professor Emeritus William Tiller quietly obliterated the conventional view that humans cannot meaningfully interact with their experiments. "Conventional science would even more emphatically state that specific human intentions could not be focused into a simple electronic device, which is then used to meaningfully influence an experiment in accord with the specific intention. We have made a valid test and have found the conventional science conclusion to be in serious error."

In his work Dr. Tiller describes the people who are capable of sustaining high-coherence intentions as "imprinters." A group of them, for example, sat around a table while putting out the intention "to activate the indwelling consciousness of the system" so that the pH of the experimental water would increase or decrease significantly compared to the control. It did. How does he explain this? The theory used by Tiller and co-researcher Walter Dibble Jr. is multidimensional. These scientists see water as a special material, "well suited for information/energy transfer from this frequency domain into our conventional domain of cognition, the physical." Regarding the factor of mental capability—whether imprinters know enough science to visualize changes in pH—Dr. Tiller said, "The unseen intelligence of the universe is an even more important factor." Later he added, "In my view, it's the spark of Spirit in the cells that gives rise to the life force."

Another scientist at that meeting, Dr. Glen Rein, points out that physicists know about the existence of energy fields with properties that are not explained by classical equations. He refers to the nonclassical fields as quantum fields. Rein's work again shows that this non-electromagnetic energy—information from the primordial vacuum of space—can be stored in water and later communicate with living cells.

Perhaps Viktor Schauberger's most startling observation was that subtle qualities of water can affect humans mentally and spiritually—influencing either the revitalization or deterioration of society. Dr. Thomas Narvaez has proved to his own satisfaction that a vitality factor exists and can be increased or decreased in water by human activity. "We now see that our thoughts not only affect our own bodies, but also the bodies of those around us. Members of this group [speaking to the Institute of Advanced Water Sciences] who bottle water or who work with broadcasted energies like crystals or magnets therefore have a responsibility to keep our view of the world upbeat and positive."

19 Madame Curie and the Spirits

What Are We to Make of the Strange Alliance between a Nobel Prize–Winning Scientist and a Notorious Medium?

John Chambers

The contrast between the medium and the female scientist in attendance at her séance could hardly have been greater.

The year was 1905, the place the Psychological Institute in Paris, France. The medium was Eusapia Palladino, the dominant European psychic of her day and the first to be examined exhaustively by many of the world's leading scientists.

The female attendee was Marie Curie, the first woman to achieve worldwide fame as a scientist, and, in 1903, cowinner with Henri Becquerel and her husband, Pierre, of the Nobel Prize in physics for their work on radioactivity. (In 1911, Marie Curie would receive a second Nobel Prize, this time in Chemistry, for discovering radium and polonium and for isolating radium.)

Eusapia Palladino, born in 1854 in the mountain village of Minervo Murges, Italy, could neither read nor write. In childhood she hit her head so badly that there was a hole in her skull, which pulsed when she was in a trance; according to the savants, the fall was responsible for her fits of hysteria, sleepwalking, epilepsy, and catalepsy. Her mother died giving birth to her, her father was murdered when she was eight, and her grandmother abused her and put her out to work as a servant girl when she was fourteen. Eusapia spoke a gutter Italian and, when in trance, a bizarre mixture of Italian and French that was almost incomprehensible. Tempestuous, usually in a towering rage, this non-educable psychic hated to bathe, loved to drink, and was constantly seducing sailors.

The world-famous scientist who held Eusapia's hand at the séances in 1905 couldn't have been more different. Marie Curie, née Manya Sklodowska, was born in Warsaw, Poland, in 1867. She was raised by loving, highly intelligent, and cultivated parents; her mother was a gifted pianist

Fig. 19.1. Born in Naples, Italy, Eusapia Palladino (1854–1918) was a prominent psychic of her day whose powers intrigued Marie Curie.

and headmistress of a girl's school, while her father was an impoverished scientist and an underinspector and teacher in a Russian government–run high school. As an adult student in Paris, Marie spoke, read, and wrote Polish, French, Russian, and German and had in-depth knowledge of other languages. She finished first in her master's degree physics program at the Sorbonne at age twenty-five and second in the math program for a second master's degree at age twenty-six; both times she was the first woman to complete the program. She completed a doctorate from the Sorbonne at age thirty-six—but almost as an afterthought, since her major scientific discoveries were behind her. Marie moved with ease among the greatest minds of her time. Though unorthodox and liberated in all her thinking, she conducted herself with propriety—except for one passionate love affair, two years after her husband's death, with the brilliant but married scientist Paul Langevin. She authored several books, including her autobiography in English.

Marie Curie was pretty; Eusapia Palladino was not. The wisp of a Polish girl had a porcelain complexion, gossamer ash blond hair, high cheekbones, and intense gray eyes that were kind when they were not lost in thought. Marie's perfect posture brought out the fetching slimness of her figure—slender ankles, slender wrists, and a very narrow waist. A certain austerity, even a grimness, came to mask her features and slow her gait as she grew older, but she never lost the delicate beauty of her physical appearance.

Eusapia Palladino, on the other hand, was physically unattractive. She was short, tended to fat, swathed herself in black, and walked with a waddle. Her mouth was twisted in a permanent downward curve that expressed—disdain? sarcasm? suffering? Nobody knew. Her eyes, sunk in a double-chinned bulldog face, crackled in their depths with a sinister fire that presaged her sudden explosions of fury. Eusapia's unchained sexuality gave

Fig. 19.2. World-famous scientist Marie Curie (1867–1934).

her a vibrant allure, but the scientists searching for occult energies in her found this easy to ignore.

What did Eusapia Palladino do? Things that made her seem to come from a different planet in comparison with Marie Curie. Deborah Blum describes her activities in *Ghost Hunters: William James and the Search for Scientific Proof of Life After Death* (Penguin, 2006): "She made furniture fly. She caused marks to appear on paper by merely extending her hand. Tied to a chair, she caused fingerprints to appear in a smooth block of clay across a room. . . . During one séance in Genoa, lights glittered overhead like dancing fireflies. One light settled on the palm of an observer, a German engineer." Eusapia made objects appear out of nowhere; she aurally channeled utterances from the spirits; she performed automatic writing; she extended her body "ectoplasmically," touching others with nonmaterial arms.

She did much more, and she did it all intermittently, unexpectedly, never under duress—and, not infrequently, fraudulently. To the end of their days, the distinguished scientists in attendance remained bitterly divided about the nature of her achievements.

Why did Marie Curie become involved with this? Deborah Blum answers the question on the op-ed page of the *New York Times* for December 30, 2006:

The scientific study of the supernatural began in the late nineteenth century, in synchrony with the age of energy. It's hardly coincidental that as traditional science began to reveal the hidden potential of nature's powers—magnetic fields, radiation, radio waves, electrical currents—paranormal researchers began to suggest that the occult operated in similar ways.

A fair number of these occult explorers were scientists who studied nature's highly charged circuits. Marie Curie, who did some of the first research into radioactive elements like uranium, attended séances to assess the powers of mediums. So did the British physicist J. J. Thomson, who demonstrated the existence of the electron in 1897. And so did Thomson's colleague, John Strutt, Lord Rayleigh, who won the 1904 Nobel Prize in Physics for his work with atmospheric gases.

Rayleigh would later become president of the British Society for Psychical Research. And he would be joined in that organization by other physicists, including the wireless radio pioneer Sir Oliver Lodge, who proposed that both telepathy and ghostly appearances were achieved through energy transmissions connecting living minds to one another and perhaps even to the dead.

We have an eyewitness account of a Eusapia Palladino séance that Marie Curie attended. (She and Pierre attended a number in 1905.) It comes from Charles Richet, Nobel Prize winner in physiology in 1913 and a leading contemporary European researcher into occult phenomena:

[The séance] . . . took place at the Psychological Institute at Paris. There were present only Mme. Curie, Mme. X., a Polish friend of hers, and P. Courtier, the secretary of the Institute. Mme. Curie was on Eusapia's left, myself on her right, Mme. X, a little farther off, taking notes, and M. Courtier still farther, at the end of the table. Courtier had arranged a double curtain behind Eusapia; the light was weak but sufficient. On the table Mme. Curie's hand holding Eusapia's could be distinctly seen, likewise mine also holding the right hand . . . We saw the curtain swell out as if pushed by some large object . . . I asked to touch it . . . I felt the resistance and seized a real hand, which I took in mine. Even through the curtain I could feel the fingers . . . I held it firmly and counted twenty-nine seconds, during all which time I had leisure to observe both of Eusapia's hands on the table, to ask Mme. Curie if she was sure of her control . . . After the twenty-nine seconds I said, "I want something more, I want *uno anello* ('a ring')." At once the hand made me feel a ring . . . It seems hard to imagine a more convincing experiment . . . In this case there was not only the materialization of a hand, but also of a ring.

What was Marie Curie's reaction to this séance? We don't know. But we do know the reaction to Eusapia Palladino's séances of her husband, Pierre Curie, a scientist whose distinguished accomplishments in the fields of piezoelectricity, symmetry in physical phenomena, magnetism, and, later, radioactivity, made him a power in his own right. Maurice Goldsmith writes: "The Curies, especially Pierre, believed in Spiritualism. . . . Pierre felt Palladino worked 'under [scientifically] controlled conditions.' After a séance at the Society for Psychical Research—where in a brightly lit room 'with no possible accomplices' he watched as tables mysteriously lifted into the air, objects flew across the room, and invisible hands pinched and caressed him—he wrote George Gouy, 'I hope we are able to convince you of the reality of the phenomena or at least some of them.'

"A few days before his death Pierre had written of his last Palladino séance, 'There is here, in my opinion, a whole domain of entirely new facts and physical states in space of which we have no conception.' In 1910, four years after Pierre's death, when Marie was rejected by the Academy of Sciences, Henri Poincaré wrote that Pierre's spirit had come to Marie and tried to comfort her by saying, 'You will be elected next time.'"

There was a moment when a belief in the afterworld seemed to burst suddenly, achingly, out of Marie Curie. This was when Pierre, whom she passionately loved, and who by every account was an exceptionally good man as well as an exceptionally good scientist, died suddenly in a traffic accident in Paris on April 19, 1906. He had slipped while absent-mindedly crossing a street in the rain; his head was crushed beneath the wheels of a heavy carriage and he died almost immediately. He was forty-seven.

Marie never came close to recovering from this loss. Some twenty-four years later, when she sat down to reconstruct a chronology of her life, she wrote that on

Fig. 19.3. Pierre Curie, noted scientist and beloved husband of Marie Curie, was tragically killed in 1906.

April 19, 1906, "I lost my beloved Pierre, and with him all hope and all support for the rest of my life." In the days following Pierre's death, she wrote in a private diary (which became public knowledge only many years later) heart-wrenching words that—albeit torn from her in a moment of awful shock—suggest her belief in the spirit world was more than passing:

"I put my head against [the coffin]," she wrote. "I spoke to you. I told you that I loved you and that I had always loved you with all my heart. . . . It seemed to me that from this cold contact of my forehead with the casket something came to me, something like a calm and an intuition that I would yet find the courage to live. Was this an illusion or was this an accumulation of energy coming from you and condensing in the closed casket which came to me . . . as an act of charity on your part?"

She added: "I sometimes have the absurd idea that you are going to come back. Didn't I have it yesterday, when hearing the sound of the front door closing, the absurd idea that it was you?"

The death of Pierre Curie was the greatest tragedy of Marie Curie's life. She bore it with a grim and bitter fortitude. She had been schooled in fortitude; tragedy was almost the lot of every Pole in Europe in the nineteenth century. After Napoleon's defeat at Waterloo in 1815, Poland had been ceded to Russia, Prussia, and Austria. Russia abolished Poland's name and for the next century sought to absorb the country into itself; the Poles would not regain their sovereignty until the end of World War I. Two unsuccessful uprisings against the Russians, carried out in 1830 and in 1863, made matters worse. The vindictive Russians did not allow higher education for Polish women; Marie, thirsty for knowledge, pursued her education in clandestine "flying" classes or by herself. Eking out enough money through governess jobs to come to Paris, she spent her years of study in an unheated garret room, subsisting on small portions of tea, chocolate, bread, and fruit and sleeping only a few hours at a time as she studied day and night. Achievement brought awards and grants her way. But her heroic years of endeavor had steeled her against adversity, even while they taught her to discount nothing and to reach out in every direction; if this hardheaded woman of genius chose to spend precious time with Eusapia Palladino, perhaps that's an invitation to us to give this unruly medium of genius at least the benefit of the doubt.

20 India's Mystic Military

Are Fleeing Tibetan Monks Changing the
Balance of Power on the Indian Subcontinent?

John Kettler

The Chinese invaded Tibet, and the Dalai Lama fled. This much many
people know. Fewer know, though, of the systematic, ongoing ravag-
ing and suppression of Tibetan culture, institutions, monuments, writings,
and, above all, the people themselves, who have been made strangers and
subhumans in their own land, deliberately dispossessed by waves of Chinese
immigrants brought in to Sinize Tibet. And the Tibetan monks have been the
direct and particular targets of Chinese repression, for it is they who iconify
the essence of the Tibetan culture, its deep spiritual beliefs that are the polar
opposite of communism, and in whom reside much of its cultural tradition.

Persecuted in their own country, many, like the Dalai Lama before them,
have fled to India, reportedly bringing with them a raft of mystical abilities,

Fig. 20.1. Tibetan monks sound great ritual horns.

abilities honed by successive generations of monks over the millennia. Now, that would be newsworthy all by itself, but instead it is merely the point of departure for the real story.

Apparently, you see, the Indian military is mining these refugees for their long guarded mystical techniques, evidently seeking a unique kind of military advantage.

Yes, this does sound decidedly "out there," but readers of *Atlantis Rising* may recall the writer's earlier "The Indian Antigravity Report: Has a Modern Technological Breakthrough Been Developed from Ancient Sources?" (in *Atlantis Rising* 51). There, what can fairly be characterized as a kind of Indian Manhattan Project was unmasked, except that India has long had the Bomb, together with a variety of delivery means. This was even more advanced—antigravity, stealth, and other technology, derived from the unlikeliest of sources: ancient Indian religious poems and stories such as the *Ramayana* and the *Vedas,* with data mining using Sanskrit scholars and Hindu clerics in addition to the usual crop of military and technical experts. Now, though, another piece of what's apparently going on has surfaced, in the form of an *India Daily* online article titled "Tibetan monks can become invisible and fly—stealth and anti-gravity reverse engineering from UFOs?" (see www.indiadaily.com/editorial/2251.asp).

Here, we are informed that not only can Tibetan monks do all sorts of amazing things, though never showing them in public, but so, too, can the Hindu hermits deep in the Himalayas. Further, the article says that fighting while in stealth mode and using antigravity were common events in the same sacred Hindu literature described above. Evidently, the Indians are trying to reconnect with an all but lost part of their full capabilities.

ANCIENT MYSTICAL CAPABILITY IN MODERN WAR

The literature of what's been seen and occasionally even filmed in Tibet by travelers is remarkable. We read of monks stripped down to loincloths draping wet towels around their necks while sitting cross-legged outside in the snow in winter and—through special breathing designed to raise the chi, or life force—competing to see which one dries his towel first. The applications of that ability alone to combat in winter and adverse weather are obvious, for cold, wet troops tend to be ineffective and frequently become casualties.

There are accounts of visitors observing monks on foot, sometimes heavily laden, blazing across the countryside in rapt concentration and at a speed more familiar to those who like to watch the Roadrunner cartoon character in action. When they have expressed a desire to stop and talk to these paragons of human performance, though, the visitors have consistently but politely been warned off, being told that such an interruption could "damage" the monk, though not the mechanism, other than that it constituted a "shock" to the monk's system.

Contrary to what we might think from the news and films, even modern combat requires a lot of walking, and there are many places where vehicles can't go at all. The British relearned this lesson the hard way in the Falklands when they suddenly found themselves deprived of helicopters after the transport *Atlantic Conveyor* was sunk, forcing them to send heavily laden, hungry troops on a grueling cross-country march across the island en route to a battle at Goose Green. Plenty of battles have been won throughout history by forces that simply outmarched their foes and seized critical ground first. What if the Indians could train their troops to get into a mental state in which they could consistently move at a pace faster than the other side could even force-march its own troops?

Modern armies tend to use a lot of heavy equipment, such as armored combat tractors, cranes, bridge-layers, and so forth to assist them in their tasks, though there is still plenty of manual labor involved in digging fighting positions, filling sandbags, building bunkers, and the like. What if, though, there was another way to get these sorts of results, but without all that gear and consequent maintenance and logistic requirements? Turns out the Tibetans have something stunning to offer—acoustic lifting and positioning of heavy objects. Yes, you read that correctly!

Antigravity researcher Bruce Cathie, in his monograph "Acoustic Levitation of Stones," part of *Antigravity and the World Grid,* edited by David Hatcher-Childress, describes how in 1939 a Swedish doctor named Jarl, while on a scholarly visit to Egypt from Oxford, was contacted by a messenger from a Tibetan friend who'd been a fellow student with him in England. The messenger bore an urgent request that Dr. Jarl go to Tibet and treat an old sick lama (Tibetan monk). The lama was important, Dr. Jarl's visit was long, and because of this, he not only was given unprecedented access to what was traditionally hidden, but was even allowed to film it! And what films! They showed carefully arrayed drummers, trumpeters, and singing, chanting monks aligned in a 90-degree arc 63 meters (206 feet) from a 1 x 1.5 meter (3.3 x 4.9 feet)* stone set in a 1 meter (3.3 feet) bowl in a flat polished stone

---

*A third dimension was not given, but it was probably 1 meter (3.3 feet), to match the bowl description.

in the meadow. The instruments amounted to thirteen drums and six Tibetan trumpets.

What happened next bears direct quotation:

When the stone was in position the monk behind the small drum gave a signal to start the concert. The small drum had a very sharp sound, and could be heard even with the other instruments making a terrible din. All the monks were singing and chanting a prayer, slowly increasing the tempo of this unbelievable noise. During the first four minutes nothing happened, then as the speed of the drumming and the noise increased, the big stone block started to rock and sway, and suddenly it took to the air with an increasing speed in the direction of the platform in front of the cave hole 250 meters high. After three minutes of ascent it landed on the platform.

Continuously they brought new blocks to the meadow, and the monks, using this method, transported 5 to 6 blocks per hour on a parabolic flight track approximately 500 meters long and 250 meters high.

For those of you not accustomed to the metric system, that's repeatedly and in a controlled manner, using nothing but concentrated sound and prayer, launching rocks weighing hundreds of pounds eighty two stories high and almost a third of a mile downrange. In this case, the rocks were being used on a high cliff ledge, accessible only from above by the monks in the work party, to wall off a natural tunnel. Just think what it would've taken otherwise, in both equipment and labor in that most out of way place, to do this job. Even today it would be a big, expensive job.

The military applications of this acoustic levitation approach were so patently obvious that the English Scientific Society, for which Dr. Jarl worked, seized the two stunning films Dr. Jarl shot, declaring them "classified." They were not to be released until 1990, and the writer doesn't know whether or not this in fact occurred.

The referenced monograph contains a much fuller description, together with a detailed discussion of the underlying special mathematics thought to make the levitation phenomenon work.

SAME OR BETTER RESULT, DIFFERENT APPROACH

One of the recurring themes in *Atlantis Rising* and in much of the writer's own delvings is that there are lots of ways to get a given effect, and it most assuredly doesn't have to be the way it's done in the West. A simple example still true today is in saw design. Western handsaws cut on the downstroke, whereas Japanese saws, having their teeth the reverse of those on our saws, cut on the upstroke, a method they deem more "harmonious" and, in an animist perspective, "respectful to the wood." Judging by their long history of magnificent wooden constructions ranging from puzzle boxes to temples, this hasn't hurt them in the slightest. Indeed, certain Japanese woodworkers' creations are heavily collected in the West, precisely because of their elegance and craftsmanship.

The above is but the microcosm of a much larger point. Most of us in the West, especially in the scientific, technical, and medical communities, tend to think that our way is the only and best way, when the evidence clearly shows that many of our "greatest discoveries" and "most fantastic achievements" are merely recapitulations of what was done hundreds, thousands, and sometimes even tens of thousands of years before. Embarrassing and upsetting, but all too true! This leads to what the Bible so eloquently calls "straining at the gnat."

Christopher Dunn has shown, for example, that even the most modern granite machining techniques could not produce the fantastic flatness and orthogonality he personally saw and measured within the granite sarcophagus in the Serapeum at Saqqara and inside the Great Pyramid at Giza. Unfazed, though, mainstream archaeology continues to hold that this was done with "mark one eyeball," copper chisels, and diorite balls. Sure! We're likewise told that a million 2.5-ton limestone blocks were moved on sledges (with or without rollers) up gently sloping ramps to build the Great Pyramid, conveniently forgetting the core of said pyramid was made from 70-ton blocks of granite!

Trepanning (cutting holes in the skull) was once thought to be a modern technique—until evidence was found that Neanderthals did it and patients survived for years afterward. The obsidian blades of the ancients turn out to have been sharper than our finest steel surgical knives, and they are now back in use as scalpels. We thought we figured out how to make and use electricity, only to discover that the ancients were electroplating base metals with gold and we now have evidence the Egyptians may have had some sort of electric lights. Western clockmakers were very impressed with themselves until the appearance of the ancient Greeks' Antikythera device, with its ability to track and predict complex astronomical phenomena via an incredibly intricate system of gears, which only recently was definitively analyzed and reconstructed.

"Okay. Okay. The ancients did some incredible stuff, but we invented the Bomb!" Sorry, late there, too, as attested not just in the *Ramayana* and the *Vedas,* but also in the very walls of Mohenjo Daro, where researchers have found not only the same terrible "human shadows" that were created at Hiroshima and Nagasaki, but also elevated radiation levels, a situation also found where Sodom and Gomorrah once stood and where longstanding tradition warns not to water the animals lest they sicken and die, verified per Zechariah Sitchin through scientific tests in the 1930s. And did you know that the correct translation from the Hebrew for the "pillar of salt" Lot's wife was turned into in the Bible should actually read a "pillar of ash," exactly what briefly happens to people close to Ground Zero when a nuclear or thermonuclear weapon detonates? This is tame compared to the wealth of explicit and even gory accounts we find in the Hindu sacred literature, with cities shattered, armies consumed in pillars of smoke, and worse. As if that's not bad enough, missiles and beam weapons of several sorts were also in use!

Impressive as all these ancient achievements are, though, even they are not the main point made in the *India Daily* article or that the writer seeks to make, which is that the more truly advanced a society becomes, the less it needs technology per se, and the more it's able to accomplish things through thought and focused will, what the article terms "spiritual power,"

Fig. 20.2. Indian soldiers test weapons at a U.S. Army facility. (Photograph courtesy of the U.S. Army)

a concept we encounter again and again in ancient texts, in discussions with shamans, and in a stack of UFO contactee reports.

We read that certain native cultures (e.g., the Hopi and the Australian Aborigines), by eschewing the advantages of modern life, act as a substantial and conscious offset to the stress and strain inflicted on the planet by the "life in the fast lane" high-tech cultures. We read that what less advanced civilizations do with technology, more advanced ones do through the powers of their minds, literally molding reality to suit their needs. Which of course goes right back to all those warnings from religious and metaphysical teachers about guarding our thoughts and our tongues.

The positive side of this can be seen in concepts such as "conscious creation" and "co-creation."

Certain sources talk about earth's being in a quarantine because its inhabitants would cause wholesale havoc in the universe beyond, which works on the principle of bringing things into being by simply speaking them. A familiar, high example of this concept is the divine decree "Let there be Light!" If we here on the planet can't, though, for whatever reason, manage to live in harmony with each other, can you imagine the incredible upheaval the unsettled minds alone could cause if unleashed upon the galaxy, let alone those deliberately choosing to throw a spanner in the galactic works?

TOWARD A HYBRID INDIAN MILITARY

India appears to be following a three-track approach. Numerically, it has one of the largest military forces on the planet. It is also growing increasingly more high tech with each passing year, a trend greatly helped by large numbers of computer- and math-savvy, well-educated people. India has a considerable and growing industrial and military manufacturing base, allowing it to produce such military sinews as nuclear weapons, computers, electronics and opto-electronics, missiles, aircraft, tanks, artillery, small arms, and ammunition domestically, while buying/leasing from other countries (Russian nuclear attack subs) or building what it needs under license (Su-27 fighters). The first two tracks alone make it formidable, but it's the third that puts it into an altogether different realm, for there we find in addition to the previously mentioned mystical capabilities, which not only include flying (personal levitation, à la the scientifically much attested David Home in the 1800s), without the need for a James Bond–style jetpack, a determined effort to harness and exploit precisely the New Energy technologies that struggle to be born in the United States. We're talking serious, well-funded work in national research facilities.

PART FIVE

ASTRONOMY
FARTHER OUT

21 Is the Big Bang Dead?

A Maverick Astronomer Challenges Reigning Theory on the Origins of the Universe

Amy Acheson

In the 1960s, astronomer Halton Arp discovered that galaxies are "born" and grow into family groups. In some cases, he could trace their genealogy through as many as four generations. This is the kind of discovery that every astronomer dreams of making. It promises to improve our understanding of the universe as much as Galileo's discoveries improved our understanding of the solar system. Arp's observations should have been celebrated and promoted. Instead, as with Galileo, his work has been rejected and ridiculed by the astronomical establishment.

In his review of Arp's book, *Quasars, Redshifts and Controversies,* the astronomer Geoffrey Burbidge described what happened to Arp after this discovery: "Arp's ranking in the 'Association of Astronomy Professionals' plunged from within the first 20 to below 200. As he continued to claim that not all galaxy redshifts were due to the expansion of the universe, his ranking dropped further.

"[In the mid 1980s] came the final blow: his whole field of research was deemed unacceptable by the telescope-allocation committee in Pasadena. Both directors [of Mount Wilson and Las Campanas, and Palomar observatories] endorsed the censure. Since Arp refused to work in a more conventional field, he was given no more telescope time. After abortive appeals all the way up to the trustees of the Carnegie Institution, he took early retirement and moved to West Germany."

Fig. 21.1. Author and astronomer Halton Arp.

What makes this discovery so important? Why was Halton Arp willing to sacrifice a promising career in astronomy in defense of it?

First, the why: Arp is one of those pioneers whose motivation is discovering how the universe works. He wants to follow the trail of this mystery until it is solved. This is more important to him than his reputation as an astronomer. Is it worth the sacrifice? Arp's wife, also an astronomer, put it this way: "If you are wrong, it doesn't make any difference; if you are right, it is enormously important."

Second, the what: Arp discovered a major flaw in one of the tools of modern cosmology. This tool, the redshift, is believed to be a Doppler shift—a measure of velocity and nothing else. Arp has proved that a large component of the redshift is intrinsic (a property of the galaxy or quasar itself), not due to velocity. In order to understand why an intrinsic redshift is such a threat to mainstream astronomers, we need to review the currently accepted theories of cosmology.

A CHAIN OF COSMOLOGICAL THEORIES

From the viewpoint of modern cosmology, there was only one event. Twelve or fifteen billion years ago, the granddaddy of all black holes exploded, creating the universe. Everything that has happened since is fallout, aftershocks, and shrapnel. The universe was given one initial burst of energy, and it has been winding down ever since. But the Big Bang isn't something we can see through a telescope. The Big Bang is a theory.

In fact, the Big Bang is part of a chain of theories. Each theory is linked to the theory next in line. The Big Bang was invented to explain how the expanding universe started expanding. The expanding universe was invented to answer the question "Why are the galaxies all moving away from each other?" The movement of galaxies is an outgrowth of the Doppler interpretation of redshift, which assumes redshift is due to movement of the light source away from the observer. Redshift is a measurement of how much the lines in the spectrum of a distant light source are shifted toward the red end of the spectrum. The observation anchoring this chain is the correlation of increasing redshift with decreasing luminosity. Therefore, the fainter (and presumably farther away) a galaxy is, the faster it is moving away. Thus the chain of theories is based on the assumption that redshift is a measure of velocity and only a measure of velocity.

For astronomers, "redshift is a measure of velocity" was a handy assumption. Combined with the correlation with decreasing luminosity, it created a yardstick they could use to determine distance. High redshift means far away; low redshift means nearby. This is useful because most of the millions of galaxies are too far away to measure their distance by any other means. The second half of the assumption, "and only a measure of velocity," was ignored.

Most of the cosmology of the twentieth century is based on this chain of theories. If any link in the chain is wrong, the entire chain collapses and we need to start over. Thousands upon thousands of professional astronomical articles, textbooks, popular magazines, doctoral theses, Internet Web pages, and press releases could become obsolete overnight. This is the threat that keeps most astronomers from looking for a flaw in the chain.

Halton Arp has found the flaw.

THE CHAIN IS BROKEN

In the 1960s, quasars were discovered. Several quasars were already known but not recognized as anything special. They were thought to be stars in our Milky Way with a few odd characteristics, such as their blue-violet color and their association with strong radio sources. Then their redshift was measured. It was much higher than that of the farthest known galaxies.

What a shock! The redshift yardstick demands that these objects are far beyond the galaxies. If that's true, how bright must they be? Astronomer Tom Van Flandern describes the problem facing astronomers: "There must exist an unknown energy mechanism to produce such intrinsically high-luminosity objects, enabling them to be so bright at such great distances. Energies are often equivalent to thousands of supernovas per year." There was no known mechanism for producing that much energy.

It would be easier to move the quasars closer. But the redshift yardstick is inflexible, so the quasars were assumed to be at the distance dictated by the chain of theories.

At the time, Arp was working on a project for which he would become famous. It was a photographic catalog of peculiar galaxies, those cosmic oddballs that don't look like the majority of galaxies. He organized them into categories—galaxies with missing or extra arms, multiple interacting galaxies, galaxies with extra-bright nuclei, disrupted galaxies, and so on. No wonder he was the first to notice that many of the newly discovered high-redshift quasars seemed to be surprisingly close to the galaxies numbered 100 through 163 in his catalog. These galaxies are the Seyfert galaxies (also called active galaxies) and the starburst galaxies.

Arp found that many of the quasars occur in pairs and lines and arcs, with a low-redshift Seyfert galaxy sitting nearby. Often the Seyfert galaxy is positioned so that its family of quasars seems to have been ejected from both ends of the Seyfert's spin axis. X-ray jets and radio lobes of the galaxy point directly at the line of quasars, often enveloping them. How could this be if the quasars are half a universe away from the Seyfert?

One or two or even a dozen quasars positioned near ordinary galaxies might be coincidence. But Seyferts are rare and spectacularly different from ordinary galaxies. They have enormous bright nuclei. Often they are bracketed by twin giant lobes of radio and X-ray material pouring out in the same directions as their families of quasars. Some, like M82, are disrupted, exploding, torn apart. Others, like M87 and CentaurusA (CenA for short), have jets that stretch thousands of light-years. And from the end of these spectacular jets, these galaxies are spraying quasars across the sky.

What does this mean? It means that quasars are neither stars nor superluminous galaxy cores. Arp explains, "Astronomers have replaced one wrong idea with another wrong idea. When restored to their rightful distance, alongside the galaxies they are ejected from, they become brighter than stars, but fainter than most galaxies." Arp thinks of them as newly created matter that will eventually become a full-sized galaxy.

Arp's discovery was that quasars and active galaxies belong together. In spite of their different redshifts, the distance is the same. This implies that the yardstick of "redshift equals distance" is mistaken. If the yardstick is wrong, the whole chain of theories based on it is wrong, including

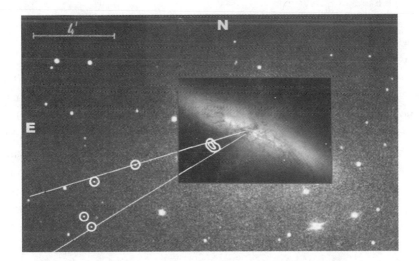

Fig. 21.2. Galaxy M82, which has X-ray knots at its core and a magnetic aura.

NGC 4258
ROSAT-PSPC
0.1-2.4 keV

5 arcmin

Fig. 21.3. M106 (also called NGC 4258) is a spectacular Seyfert with a pair of what may be quasars flanking both sides of its active nucleus.

the Big Bang and expanding universe. That means it's time to look for another cause of redshift in quasars. It's time to weed out the distortions caused by arranging the universe according to this faulty yardstick.

Once he recognized that these galaxy-quasar families contradicted the redshift yardstick, Arp turned his telescope to other objects whose distance was determined by redshift. He discovered that the distances of galaxies and clusters of galaxies were also warped, in specific quantized jumps, by the faulty yardstick. Even the stars in our Milky Way show small non-velocity redshifts.

How will these observations affect our image of the universe? The Big Bang theory predicts when the universe was born, its size and shape, how galaxies evolve, how the universe will end. All these characteristics will change in Arp's universe.

Let's compare the two cosmologies.

HALTON ARP'S UNIVERSE

In Arp's cosmology, a large component of redshift means age, not velocity. The higher the redshift, the younger the galaxy or quasar. Galaxies that have no excess redshift are the same age as the Milky Way. The seven galaxies (out of millions) that are blueshifted are older than the Milky Way. Six of these blueshifted galaxies are in the Virgo Cluster and the seventh is M31 (also known as the Great Nebula of Andromeda), our nearest neighbor in space and the dominant galaxy of our local group of twenty to thirty galaxies.

1. How was the universe born? According to the Big Bang theory, the universe exploded into existence from nothing twelve to fifteen billion years ago. In Arp's universe, the same data point to a different

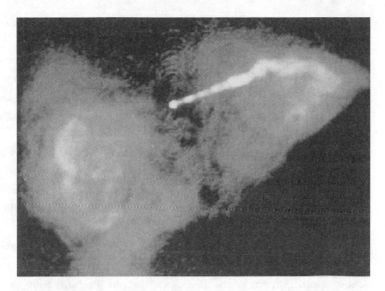

Fig. 21.4. *CentaurusA in optical light with jet in X-ray. (X-ray image provided courtesy of NASA/CXC/SAO. Optical image provided courtesy of URA/NOAO/NSF)*

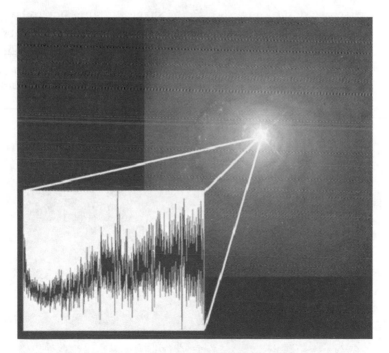

Fig. 21.5. *The Chandra X-ray telescope focused on a blueshifted cloud at the center of NGC 5548, completely ignoring a nearby high-redshift galaxy cluster.*

event twelve to fifteen billion years ago—the "birth" or ejection of the Milky Way.

In Arp's universe, M31 is the parent galaxy of our Milky Way. Twelve to fifteen billion years ago, M31 was an active galaxy and the Milky Way was a knot of plasma in M31's jet. Unlike the slanted jet of CenA (see figure 21.4), M31's jet was aligned straight down the spin axis. How do we know this? M31 isn't active today—it no longer has a jet. But we can deduce the direction of M31's jet from looking at M31's family, our local group of galaxies. After billions of years, this family is still lined up in a remarkably straight line.

The Milky Way is a parent as well. The Magellanic Clouds, visible from the Southern Hemisphere, are two of its offspring.

At a recent workshop, I asked Arp what happens when galaxies grow old. He replied, "We don't have enough information yet to know what happens when galaxies grow old. Perhaps they exhaust themselves and fade away."

2.   How big is the universe? According to the Big Bang, the universe is a sphere about thirty billion light-years in radius. In Arp's universe, we don't know the size or shape for sure, not today, and not billions of years ago. All we can know is that it stretches out in every direction farther than we can see. It is also older than we know, possibly infinite and eternal. But for that part of the uni-

Fig. 21.6. M87 *with jet. (Image courtesy of Merlin)*

verse we see in our telescopes, high-redshift objects are closer than the redshift yardstick indicates.

Many modern astronomical concepts, such as curved space-time, are really only attempts to compensate for the distortion of measuring the universe with a broken ruler. Arp puts it this way: The universe is "flat and Euclidean. Gone is the logic defying curved space, which so many people have strained to imagine (not to mention curved time). Gone are the hypothetical singularities [black holes] where physics just breaks down; and gone, too, is a universe made up of more than 90 percent invisible matter."

3. How do galaxies evolve? According to the Big Bang, scraps of matter from the initial explosion fall together under the influence of gravity to make a galaxy. In Arp's universe, active galaxies eject new matter in the form of high-redshift quasars. The quasars gain mass and lose redshift in even jumps as they grow into mature galaxies. We can watch the process today. Of course, press releases on the eleven o'clock news won't mention Arp's interpretation, but if you know what to look for, you can see it yourself. Galaxies don't "blow bubbles"—they eject quasars. Black holes don't come in standard sizes—they don't even exist. Arp explains, "Black holes, where everything is supposed to be falling in, are a poor explanation for the cores of active galaxies, where everything appears to be falling out."

4. How does the universe end? According to the Big Bang, there are three possibilities: It might continue to expand forever. Or it might reach equilibrium and remain stable forever. Or the expansion might stop and all the galaxies would collapse back into a black hole. None of these scenarios applies to Arp's universe. We really have no idea how it will end or if it ever will. That's an open question, a mystery for future researchers to solve.

5. When the distortions of the redshift yardstick are removed, a big picture of the form of the universe comes into focus. The whole population of the sky becomes two enormous spiral superclusters. Our local group lies between them, possibly along the arms of the brightest supercluster. These superclusters are centered on the Virgo Cluster in one part of our sky and the Fornax Cluster on the other. Spirals of spirals. Galaxies of galaxies. And who knows what lies beyond, waiting to be discovered? Halton Arp gives us a hint: "It might be something very big."

22 The Cycles of Danger

Does New Research Mean We Are Headed for Trouble?

William Hamilton III

We are conditioned to look at history as a linear progression from one stage of development to the next with irregular gaps in the time-line to be filled in as our database grows. The generally accepted picture from mainstream science is of a gradual evolutionary development of the geosphere and biosphere up to the present day. Any hint that there is some anomaly in this progression is met with skepticism from the scientific community, an attitude that is all too familiar.

But most of you have realized that when confronted with stories, tales, and epics of bygone eras, there is a need to consider that they may represent extraordinary evidence of true events, not just mythological fabrications invented by our creative ancestors. Thus, we should treat the ancient tales of pre-diluvian civilization as something based on fact, worthy of investigation and exploration rather than dismissal. Such accounts as the tale of Atlantis are seen as potentially real, not merely symbolic. In these stories we hear, time and again, of cataclysm and catastrophic change that have altered the geological record and wiped out the substantial evidence that we seek, but should we believe it?

Despite the orthodox penchant for thinking of time as a linear progression of events, plenty of evidence that natural movements do indeed occur in cycles has turned up—just as the priests of Sais told Solon in Plato's account of Atlantis. And we should not be surprised. After all, when one looks at the structure of the cosmos from the atomic to the galactic, a pattern of motion in circles and spirals is apparent, a regular repetition of conditions as in the recurring seasons of the year and the motion of the planets through the zodiac.

If a celestial body, in an ongoing repetition of orbits, returns to a previous position with respect to other celestial bodies, we have a cycle, and the possibility for encounters with previous conditions that exist relative to the planet's location in space. These cycles can be investigated for repeating events that could signal possible catastrophic changes and, perhaps, even the periodic extinction of life.

In fact, an analysis of the past reveals both small and large cycles of time to consider. The purpose in modeling these cycles here is to see if it might be possible to predict future threat zones and thus to prepare for what could be catastrophic changes to our world.

Let us briefly look at one much talked about instance that some say resulted in the destruction of Atlantis. This event, if it recurs as predicted, is just a few years from now.

From time to time a white dwarf star will accumulate too much hydrogen gas from a neighbor, a process that ultimately eventuates in a tremendous explosion of this gas shell that lights up the star in the heavens. This is what we call a nova. It usually occurs at the final stages of a star's life cycle.

Do we know all that we need to know about novas? What would happen, for instance, if a cloud of hydrogen gas of unusually high density were to engulf our sun? Could a mini-nova result in the expulsion of a shell of gas that would burst like a firestorm through the solar system? Although it seems unlikely, studies of ancient history indicate variations in solar output that may have produced catastrophic changes on earth. Even today, a variation in solar luminosity is occurring and scientists report that the slight increase in solar output may be contributing to climate change and global warming. There is also evidence that some of the other planets in our system are experiencing warmer temperatures and climate change too. These changes could be the result of increasing accumulations of cosmic dust through which our solar system is passing.

This writer's interest in the sun has recently been stimulated by reports received from a Dr. Dan B. C. Burisch, who claims he is a microbiologist who worked for a shadowy arm of the government. He tells me that preparations are being made for a catastrophe in the year 2012 that will involve changes in our sun and its effects on the earth. This is, of course, related to deciphering the Mayan symbols that are believed to point to the winter solstice of 2012.

To summarize the predictions, it can be said that a recurring event may cause a significant change in our sun. That event, known as the "grand crossing," is synchronized with a phenomenon frequently written about in these pages, the precession of the equinoxes. Many do not think that anything special will happen, but others believe that the Maya recorded significant events and used precise calendars to forecast the recurrence of periodic cycles marked by such events.

Why would the intersection of our sun and solar system with the Milky Way's equatorial plane constitute a noteworthy event? According to one website, "The auspicious year of 2012 indicated in the long-count calendar illuminates the fact that the Precessional movement of the Winter Solstice Sun will gradually bring its position into alignment with the very center of our Galaxy. For the Maya, this is like the last stroke of midnight on New Year's Eve only in 2012, the New Year, is the New Galactic Year of 26,000 solar years. The Galactic Clock will be at zero point and a New Precessional Cycle will begin" (see www.kamakala.com/2012.htm).

Closer observation of weather and climate changes seems warranted. Will the sun become more active? Indeed, there have been more X-class solar flares and coronal mass ejections in the past few years, and strangely, during a period of time when our sun, according to conventional wisdom, should have been quieting down.

The sun is just one of hundreds of billions of stars in our galaxy. The Milky Way, as we call it, is composed of gaseous interstellar media, neutral or ionized, sometimes concentrated into dense gas clouds made up of atoms, molecules, and dust. All of the matter—gas, dust, and stars—rotates around a central axis perpendicular to the galactic plane. It is generally believed that the centrifugal force caused by this rotation balances out the gravitational forces, which draw all the matter toward the center.

According to the conventional wisdom, all stars in the galaxy rotate around a galactic center but not with the same period. Stars at the center have a shorter period than those farther out. The sun is located in the outer part of the galaxy. The speed of the solar system due to the galactic rotation, it is said, is about 220 kilometers (137 miles) a second. The disk of stars in the Milky Way is about 100,000 light-years across and the sun is located about 30,000 light-years from the galactic center. Based on a distance of 30,000 light-years and a speed of 220 km/s, the sun orbits around the center of the Milky Way once every 225 million years. That period of time is called a cosmic year. The sun, it is generally believed, has orbited the galaxy more than twenty times during its 5-billion-year lifetime. The motions of the period can be determined, it is argued, by measuring the positions of lines in the galaxy light spectra.

Now, after a painstaking computer study of fossil records going back for more than 500 million years, two UC Berkeley scientists say that life on earth has flourished and then virtually vanished in cycles of mass extinction, with surprising and mysterious regularity every 62 million years.

The findings are certain to generate a renewed burst of speculation among students of the history and evolution of life. Each period of abundant life and each mass extinction, it is argued, has itself covered at least a few million years—and the trend of biodiversity has been rising steadily ever since the last mass extinction, about 65 million years ago, when dinosaurs and millions of other life-forms disappeared.

The Berkeley researchers are physicists, not biologists or geologists or paleontologists, but they have analyzed the most exhaustive compendium of fossil records that exists—data that cover the first and last known appearances of no fewer than 36,380 separate marine genera, including millions of species that once thrived in the world's seas, later disappeared, and, in many cases, returned.

A new book by Michael J. Benton, *When Life Nearly Died,* discusses the end-Permian extinction, which is believed to have happened 251 million years ago, in an attempt to understand the cause of an episode where 90 percent of life on earth was wiped out, the biggest extinction event in earth's history, much larger than the one that is said to have wiped out the dinosaurs.

Based on the understanding that our sun and its family of planets do not follow a flat elliptical path around the Milky Way, but rather an elliptical sinusoidal path, which would have the effect of lengthening the so-called

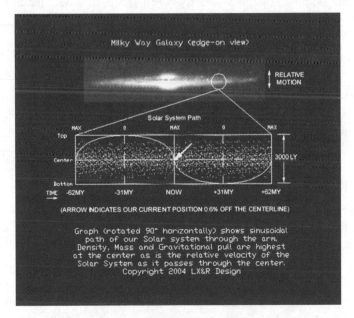

Fig. 22.1. *This illustration graphically illustrates the qualities of density, mass, and gravitational pull found in our Milky Way galaxy. (Diagram courtesy of William Hamilton III)*

cosmic year, it can be argued that the length of the cosmic year should be recalculated. Since we do not know the length of the extreme nodes of this orbital sine wave, such a calculation can be only approximate. But according to this line of reasoning, the cosmic year could be 248 to 251 million years in duration, which suggests we may be returning to the same point in our galactic orbit that we reached during a previous (extinction-level) event and that we may have entered a threat zone.

Among the considerations in this analysis should be variations in solar output likely to result in climate change; the rising probability of encounters with comets, asteroids, and dust; and the possibility of plagues resulting from bacteria and viruses embedded in dust, comets, and asteroids that could bring us diseases from space as hypothesized by the late Fred Hoyle in his book, with Chandra Wrickramasinghe, *Diseases from Space*. Another periodic event is the reversal of polarity in the terrestrial magnetic field. There seems to be some corroboration for this possible cosmic scenario in speculative periodic events such as those suggested by the Mayan calendar-based forecasts for the winter solstice of 2012.

A complete sinusoidal cycle would cross the galactic equator every 62 or 62.5 million years, with a half period of 31 million years. This is according to data obtained by this writer from researcher Bob Alexander, who has been investigating the same hypothesis, but who has used slightly different figures to make his calculation. Alexander concedes that the minor axis of the suggested elliptical wave may be less than the 3,000 light-years that he initially calculated was a major axis.

This writer used a calculator to determine the length of the perimeter of an ellipse, which involves a close approximation, using the semimajor axis and semiminor axis to determine the actual path length of the orbit based on Alexander's figures and thus the time it takes for one circuit around the galaxy.

Within a margin of error, it was determined that the entire length of the sinusoidal path would be 180,941.0769 light-years (slightly less if the semi-minor axis is shorter). This would result in a cosmic year period of 249,698,580 years; if shortened slightly, from errors in estimates, to 248 million years, we would have the following calculation: 248MY/62MY = 4.

Since, according to the Berkeley research, we have had catastrophic events approximately every 62 million years (sometimes more and sometimes less), then it follows that we may have at least four critical points where our sun and planets, components of the Orion arm of the Milky Way, might cross a critical zone that has a higher concentration of dust, debris, asteroids, and maybe even radiation that could constitute a hazard on our celestial journey. That is not all, though.

Since the earth has endured cataclysmic events on a much more frequent basis than every few million years, it may be that these horrendous events should be associated with another cycle: the precession of the equinoxes. This 25,770-year cycle, it has been argued (see chapter 7, "Precession Paradox"), could mean that our sun has a dark star companion that has an orbital period of thousands of years, which may affect our sun and family of planets so as to disrupt the space weather cycle.

We are, indeed, seeing signs of change on our sun and on earth and also on Mars—where global warming is also occurring—as well as signs of change on some of the other planets. Does this presage another cataclysmic event in our future? At this time, none of us can say for certain, but it seems possible that the examination of these cycles, in search of a meaningful pattern, could lead to the forecasting of major changes that might threaten the continued existence of civilization as we know it.

One last note of significance that deserves further scrutiny is whether recent changes in the earth's declining magnetic field and polar wandering reflect a change in the earth's core and mantle that is producing internal rising temperatures, which, in turn, are precipitating increased volcanic activity and earthquakes.

Open University researchers have uncovered startling new evidence about the possibility for an extreme period of a sudden, fatal dose of global warming some 180 million years ago. The findings could provide vital clues to the climate change happening today and that may come in the future.

For the last three to four years, the European Space Agency's Envisat satellite has been continuously monitoring the earth and gathering invaluable information for humanity. It is clear that we need to engage more space surveillance of our planet and get a better estimate on threats to life and civilization.

It's not too late for a wake-up call.

PART SIX

MEDICINE OF
ANOTHER KIND

23 Healing Vibes

Dr. Richard Gerber Is Still Convinced There's More to Getting Well than the Medical Establishment Can Comprehend

Cynthia Logan

r. Richard Gerber has just finished the half-hour commute from his two-story home in the Detroit suburbs, where he enjoyed a vegetarian breakfast on his deck, which overlooks a large garden, beautiful landscaping, and the peaceful flow and gentle sounds of a nearby river. Well nourished in body, mind, and spirit, he is now sitting in his office at St. John Macomb Hospital, just outside Livonia, Michigan, where he is a staff internist. He arrives early each day, not to study charts, but to pray for and send healing energy to each of his patients. It's a ritual he considers to be part of his job as a healer. Tuning in to his heart, he changes focus from person to person, "holding" the patient in white or pink light. It's just the kind of thing you might expect from the physician of the future, and certainly from the guy who wrote *Vibrational Medicine* back in 1988.

With its publication, Gerber provided a structure—a bridge, if you will, between the medical establishment and the alternative, complementary healing community—the "transitional scientific model that would close the gap between physics and the metaphysical." With a bachelor of science in zoology from the University of Michigan and a medical degree from Wayne State University School of Medicine, he had the credentials to support what some might consider "far-out" theories espoused in the book (now available on the CD series *Exploring Vibrational Medicine* from Sounds True). Both the book and the CDs cite numerous scientific studies that have been conducted on Kirlian photography, homeopathy, crystal healing, sound therapies, Bach flower remedies, radionics, and a plethora of other cutting-edge New Age healing modalities. Though a bit wordy, Gerber's pleasant voice and measured delivery make the CD set an informative, enjoyable introduction.

Once "very left-brained," Gerber has been interested in science since he was eleven. Yet during medical school, he began to wonder whether there were less invasive, less toxic, less costly alternatives to many of the treatments

166

he was learning about in 1976. Encountering *A Course in Miracles,* he began regular attendance at a weekly gathering for professionals interested in such concepts. Inspired, Gerber undertook an eleven-year personal study of alternative medicine; he then put together his findings to create the best seller that has been used as a textbook in various disciplines, has been translated into many languages, and continues to be the definitive book on the subject. Though he's "a bit surprised" at the book's wide acceptance and continued popularity, he recognizes his contribution. "I have added new insights to existing studies," he claims. "I have achieved a new synthesis of understanding in a field that badly needs some type of theoretical foundation upon which to build a new science of healing."

Fig. 23.1. Dr. Richard Gerber, author of the groundbreaking book Vibrational Medicine.

That new science is built on what Gerber calls an "Einsteinian model" of healing, versus the current "Newtonian paradigm." Vibrational healing philosophies consider the body more than an exquisite machine and human beings more than flesh and blood, proteins, fats, and nucleic acids. The life force, a subtle energy not fully grasped by most scientists or physicians, is the key component in the new paradigm, which sees people as networks of complex energy fields that interface with physical/cellular systems. What we're talking about here is oscillating, magneto-electrical energy that moves faster than light and is the very nature of a multidimensional universe. "It took nearly a hundred years for physicists to catch on to Einstein's profound revelations about the relationship between matter and energy," notes Gerber. "Perhaps these next hundred years will find the biologists finally catching up, integrating these same Einsteinian insights into the 'future medicine' we all hope will come about."

Gerber considers vibrational medicine to be a subspecialty of medicine based on two key concepts: we are more than physical bodies, and there are spiritual beings seeking to assist us in learning about our own multidimensionality and that of the universe. Gerber thinks it won't be too long before multidimensional human anatomy will be taught in medical schools. "We're already in the process of writing the textbooks for this," he says. According to Gerber, we've evolved the mechanical analogies of the body to where we now see it as an electronic bio-computer, which, in his view, is closer to the truth but still falls short. He won't be satisfied until counterparts to physiological control systems like the CNS and autonomic nervous systems and the endocrine system are recognized. Counterparts like the acupuncture system, whose series of points—energy pores, if you will—take in and distribute chi through the meridians, which amount to an electrical circuit board that interfaces between the physical and etheric bodies.

The etheric body is another counterpart Gerber wants acknowledged.

He feels (and cites studies, of course) that this subtle body is a holographic energy template superimposed over the physical body that guides physical cellular growth. It also, he says, carries spatial information on how the developing fetus will unfold in utero, and confers structural data for growth and repair of the adult organism (this is the means by which a salamander grows a new limb when one is severed). Gerber's elegant theory began when he started medical school, but picked up speed during his fourth year, when he worked at the Edgar Cayce A.R.E. Clinic in Arizona, and with Dr. Norman Shealy (the physician now well known for his collaboration with medical intuitive Carolyn Myss), completing an elective called Comparisons in Primary Health Care. His paper, contrasting the holistic approach with the conventional, became the nugget that later grew "in a single gestalt" into the outline that became *Vibrational Medicine.*

In 1984, Gerber grabbed one of the first Macintosh computers and began writing. Though it seemed silly to him at first, he followed his intuition and taped a quartz crystal to his third eye with a Band-Aid. "Surprisingly, this procedure seemed to accelerate the writing process," he says. Writing in the evenings after work, he had nine chapters within nine months, but the book didn't take definitive shape until he had taken a tour to Egypt. As the "trip physician," Gerber employed laying-on-of-hands healing for tour members' minor health problems, "after our supplies of Lomotil and antibiotics were exhausted."

Gerber and colleagues like doctors Andrew Weil, Bernie Siegel, Christiane Northrup, and many others offer a balanced view that may well bring in the medicine of the future. Insisting on scientific rigor and clinical studies without bias, these pioneers hold out the possibility of the recovery of ancient wisdom and its potential marriage with ever evolving technology. "In the future, homeopathic remedies and flower essences may be recognized as useful for treating various chronic ailments," writes Gerber, "but I will still rely on a good vascular surgeon to treat a ruptured aortic aneurysm." He also praises the benefits of modern medicine in eradicating epidemic diseases with vaccinations, in increasing life spans with improved hygiene and bacteriological control, in saving lives with organ transplants. He even condones the use of chemotherapy for cancers like Hodgkin's disease and childhood leukemias, and approves the use of drugs for diabetes and hypertension. He points out the already accepted uses of energy medicine, such as TENS (transcutaneous nerve stimulators) units for pain relief (and for cranio-electro stimulation to release endorphins in the treatment of depression), pulsed electromagnetic fields to stimulate new bone cells, full spectrum light to treat seasonal affective disorder, shockwaves to shatter kidney and gallstones without surgery, and music therapy in pre- and postoperative care, not to mention the latest PET (Positron Emission Tomography) scanners that show physiological and cellular function in addition to bone and soft-tissue structure.

"Western technology has now evolved to the point that we are beginning to get confirmation that subtle energy systems do exist and that they influence the physiologic behavior of cellular systems," says Gerber, who envisions "a kind of Mayo Clinic of healing research" that would be staffed by doctors, nurses, medical researchers, acupuncturists, healers, herbalists, clairvoyant diagnosticians, engineers, chemists, physicists, and others. "There would be an interdisciplinary team which could design experiments to measure the subtle energies of human function and observe how they are affected by different modalities of healing," he explains. There would also be affiliated clinical treatment centers that would have access to ongoing computerized files of healing studies and a publication similar to the *American Journal of Medicine* that would provide quotable references, eliminating what Gerber calls "the Catch-22" of current healing research (establishment publications won't entertain unconventional studies, preventing them from gaining the credibility they need to become quotable sources). Gerber would like to see practitioners teaching and healing each other. "As various modalities of healing were shown to be effective in limited studies, clinical trials would begin in a larger outreach," he suggests.

This is a dream he's working hard to manifest. He finds hope in the fact that science has begun to confirm that love is indeed a healing energy that can produce measurable healing effects. "If we can produce a ripple effect of healing energy upon the waters of humanity's collective unconscious, it will be carried by the flowing stream of earth's magnetic field and grid-work systems," he writes. In such manner, Gerber thinks we might build an energetic tidal wave of healing energy that, fueled by the power of unconditional love, could transform our planet. While he's optimistic, he's also well aware of the challenges in transforming ourselves as individuals before broadcasting healing energy on a global scale. Brightly, he brings up the theory of dissipative structures: "We only need to transform a percentage of the whole in order to effect a dynamic shift of the whole system. As more and more healers are trained, we may gradually approach the needed critical mass."

Those healers may be trained in not only conventional medicine, but also all types of nonconventional healing techniques, including Therapeutic Touch and psychic healing. Gerber would like to see an integrated approach, for example, in the treatment of heart disease. The orthodox method utilizes drugs, surgery, and angioplasty techniques to improve function. Holistic practitioners offer chelation therapy, along with visualization and stress reduction. Vibrational practitioners, says Gerber, would deal with the subtle energetic predisposing factors (such as impaired heart chakra function) and would include modalities such as flower essences, gem elixirs, and crystal therapies, along with homeopathic remedies and meridian-balancing treatments. Vibrational medicines such as flower essences and gem elixirs, and homeopathic remedies, derived from biological and mineral sources, utilize the energetic storage properties of water to transfer a frequency-specific, information-bearing quantum of subtle energy to the patient. Flowers, in particular, contain the life force of the plant and, when prepared using sunshine as part of the process, actually transfer an aspect of this life force to the remedy. Additionally, vibrational healers would offer spiritually directed counseling.

Gerber, who is married to a clairvoyant and has neither children nor pets, can afford to focus his energies on his spiritual development, his career, and his contribution to the future of medicine. What he hasn't yet articulated is a standard by which practitioners of vibrational medicine would be measured (we are, after all, talking about crystal healers, chakra balancers, pendulum swingers, past-life regression therapists, and all manner of subjective disciplines where, let's face it, charlatans abound and well-meaning people can make mistakes while making a buck). Although the future isn't clear, it is more than likely that doctors of the future will come face-to-face with their own multidimensionality and that of their patients, and will no doubt seek methods to address illness and promote wellness at every level. And, no doubt, Dr. Richard Gerber will remain in the vanguard, praying for and visualizing that very thing.

[Editor's note: Dr. Gerber passed away in June 2007.]

24 The Malady in Heart Medicine

A Doctor Shatters the Myths behind Popular Treatments for Heart Disease

Cynthia Logan

r. Charles T. McGee may be soft-spoken, but his book *Heart Frauds* gives voice to a powerful challenge that, so far, has yet to be answered. Subtitled *Uncovering the Biggest Health Scam in History*, it exposes what McGee claims are unnecessary procedures that thousands of Americans undergo each year, often under a doctor's "threat" that, without such a procedure, death may be imminent.

Though it contains numerous cartoons (drawn by a political satirist who is a former patient), and though McGee's innate humor creeps in on most pages, the book is far from funny. A board-certified obstetrician/gynecologist, McGee now finds himself delivering a potent message that may help birth a new movement in health care. Admitting he is an outsider when it comes to cardiac medicine, the research is nonetheless impeccable and remains unrefuted. The way he sees it, coronary specialists simply haven't done their homework or are ignoring the facts.

Fig. 24.1. Board-certified obstetrician/ gynecologist Charles McGee is outspoken in his belief that many cardiac surgeries are unnecessary and unwarranted.

Having read studies published in medical journals, which physicians consider their "Bible," as well as having analyzed results from low-cost, alternative methods, McGee is confident in and passionate about his accusations. "Recommending expensive, high-risk procedures over cheaper, more

171

effective ones amounts to nothing more than fraud," he states, adding that "for most people, angiograms, coronary bypass surgery, balloon angioplasty or cholesterol-lowering drugs are completely unnecessary; if your doctor recommends these, your best course of action may be to run out the door.

A "military brat," McGee grew up in San Francisco during World War II, then moved to Seattle, where he went to the University of Washington, subsequently attending Chicago's Northwestern University medical school, completing his internship and residency in 1965 in Oakland, California. Married by then, he and his wife, Carole, took vacations to Mexico, which eventually led to a year in Ecuador with Project Hope, a Navy hospital ship that would dock in a country for a year and train local doctors before moving on. It was in the Andes that he observed a culture without the diseases rampant in more "civilized" societies. "Of 330,000 Indians," he notes, "only forty to fifty would be in the hospital at one time, and then only for complications in pregnancy or as a result of traumatic accidents."

The experience piqued his interest; later, working with the CDC (Centers for Disease Control) in Atlanta as an epidemic intelligence officer, McGee became aware of biostatistics and the work of British epidemiologist T. L. Cleave, M.D. Cleave studied data from hundreds of locations around the world and did not find a single exception to this pattern: More "primitive" people eat fresh, whole foods and little or no refined sugars, refined carbohydrates, or processed foods. As soon as such foods enter a cultural diet, degenerative diseases are introduced.

After practicing OB/GYN at Kaiser Foundation Hospital in Walnut Creek, California, for a number of years, McGee began to study Chinese medicine. By 1978, his practice had become completely alternative and included homeopathy, nutrition, acupuncture, and chelation therapy. In the meantime, his father had died of a heart attack just when McGee had finished his residency, and he began a personal crusade to understand why.

One of the most interesting findings McGee reports is that cholesterol isn't the huge heart disease determinant we've been told it is. "In fact," he states, "the theory has developed a life of its own and has risen to the level of a religious dogma." He cites early animal studies that launched the theory as having been flawed, and says that "reports of later studies in humans have been slanted, and that statistics have been misinterpreted to support the theory since 1970."

According to McGee, cholesterol-lowering drugs don't lower levels of lipoprotein, which may turn out to be the most important blood test factor in determining risk of a heart attack. Furthermore, he claims that blood cholesterol levels are ineffective in determining heart attack risk and that more accurate measurements exist but are seldom used.

Now highly interested in food as medicine, McGee feels that "we're eventually going to find that food has energetic properties in addition to nutritive value." He points out that LDL, or "bad" cholesterol, is perfectly normal when not oxidized, and is native to the body in that state, whereas oxidized LDL is highly toxic. An ordinary lab isn't set up to divide the LDL into the two forms, which would be the appropriate measurement to take.

Another myth, says McGee, is that angiograms are the "gold standard."

"First of all," he explains, "there are two distinct tests, one of which is highly inaccurate—that's the one commonly administered and chances are high that, should you need one, you won't be advised of the difference." The more effective quantitative angiogram uses two cameras to achieve a three-dimensional effect and views coronary arteries from two angles simultaneously. "As of 1994, there were only twenty of the new angiogram machines in the entire world," says McGee; "even now, probably less than 5 percent of doctors even know about it!" Besides being invasive (during an angiogram, a catheter is passed up the aorta to a point just above the heart; a dye-like material is then injected, which flows through the arteries of the heart and shows up on an X-ray), interpretation of the results can vary radically.

Something that does appear to be a determinant is what is known as the "oxidation theory," where various fats pass from blood into arterial walls; if levels of antioxidants are low enough, the fats are oxidized, go rancid, and create "rust" (oxygen molecules attached to fat molecules), forming the familiar plaque. Then, according to the "rupture theory" (one of the latest developments in cardiac medicine), a heart attack occurs when the surface area over a plaque cracks. A bump grows on the arterial wall, stretches and breaks, splits open, bleeds a little, forms a clot, and causes heart muscle cells to die. The oxidative process begins with some sort of injury to the inner lining of an artery. Cells in the arterial wall immediately below this damaged area begin to actively pick up oxidized lipoprotein from the blood. Other fatty materials become incorporated into the buildup and form "atheromas," or fatty streaks. Arterial walls contain a layer of strong, circular muscles, leaving the growing streaks only one way to expand—by protruding into the opening of an artery. Over many years, the artery is gradually obstructed. According to McGee's research, "most ruptures occur in small arteries and only affect small areas of the heart . . . these small arteries don't show up even on the 3D angiograms. Physicians almost always look for obstructions in the larger arteries, when the problem is down in the smaller arteries."

Though certainly critical of the medical establishment, McGee is not radically anti-allopathic. As a physician himself, he acknowledges that "medical management of coronary artery disease has its bright spots." He cites the development of CCUs (Coronary Care Units) in hospitals as life-saving, and says that if he were to have a heart attack, he would want to be rushed to one himself.

He also acknowledges the progress of modern medicine in the area of contagious disease control, congenital defects, infections, and other acute medical conditions. His big argument is that medicine fails when it attempts to treat chronic degenerative diseases with the same approaches it uses successfully in treating acute problems. He's particularly concerned because "of all modern maladies brought to the attention of doctors, 80 percent are chronic."

And don't get him started on angioplasty "balloon" surgeries or the coronary artery bypass grafts (CABGs), which he less-than-fondly terms "cabbages." He lambastes both procedures, and cites Dr. Eugene Braunwald, professor of medicine at Harvard University, who stated that doctors should recommend surgery only when chest pains can't be controlled through other means and who predicted that a financial empire would develop around surgical procedures on the heart.

In contrast, European surgeons are paid a salary, taking the temptation to "overcut" out of the equation; in America, powerful vested interests make it hard for the system to change. According to McGee, modern medicine has never learned to monitor itself adequately enough to protect the public. He particularly questions this as it pertains to cardiac surgery. "Insurance companies already know which surgeons have high complication rates and which ones don't," he points out.

"When it comes to coronary artery disease," writes McGee, "treatments would be almost laughable if many were not so risky, costly, traumatic, and ineffective. Treatment decisions are usually made on the basis of inaccurate angiograms." Only three scientific studies have been done to determine the efficacy of bypass surgery; one, known as the Veterans study, divided cardiac patients into two groups: the first group was treated with medication, the second with bypass surgery. A follow-up study found that after ten years, the bypass group hadn't fared any better. The other two studies showed the same results.

The balloon angioplasty has become a very popular procedure despite the fact that no long-term survival studies have been conducted. In a recent short-term study, patients with coronary artery disease did just as well taking an aspirin a day as having a "balloon job."

"Heart disease victims should insist on having either an echocardiogram or a nuclear medicine isotope scan to determine the ejection fraction, a measurement of how well the left ventricle of the heart is functioning as a pump," says McGee. "If it is pumping blood normally, there is no evidence that a cabbage or balloon procedure is going to improve chances of survival." If a patient does survive a heart attack and still has chest pains that can't be controlled with drugs, he or she may be a candidate for surgery to try to relieve the pain. "But," says McGee, "there are other options. One is to load up on antioxidants and follow Dean Ornish's program, which has been accepted by health experts as proving the disease can be reversed in a high percentage of heart patients." He admits that the program isn't for everyone, but feels adamant that patients have the right to know about it and be offered a choice.

Another choice might be chelation therapy, in which a solution of EDTA (ethylene-diamine-tetra-acetic acid) is administered intravenously, drawing metabolic wastes from the bloodstream. Chelation therapy has been used on more than 500,000 patients in the United States for the past forty years, and has shown great success in managing heart disease, but the FDA has yet to grant EDTA full approval, giving mainstream physicians fuel for their fierce criticism. McGee responds with his usual factual humor: "Doctors who terrify potential EDTA chelation patients by telling them they will die should be reminded of an interesting fact: When donor hearts for transplant surgery are moved from one city to another in little coolers, they are immersed in a 100 percent solution of EDTA."

It's these kinds of facts that make McGee's book a real eye-opener. Originally published in 1994, the book was reissued in 2001 by Healthwise Publications. He is also the author of *How to Survive Modern Technology, Miracle Healing from China . . . Qigong,* and *Healing Energies.* Like the late Dr. Robert Mendelsohn (*Confessions of a Medical Heretic*), McGee has the courage of his convictions and isn't afraid to "put them out there," though he doesn't plan to "take on" the medical industry—he's seen a number of alternative practitioners be dragged into litigation and lose their licenses.

His recently published book, *Healing Energies of Heat and Light,* details his current interest in what he terms "a quantum leap in health care." He applauds the small percentage of cardiologists who have replaced their surgical practices with chelation therapy, and advises patients of any doctor to research alternative therapies for themselves. "You need to be informed that there may be options in treating a problem; however, you will need to educate yourself in this area because few physicians know about other options and therefore none will be offered. Only by being informed of all your options can a truly informed consent be granted for a procedure."

25 Energy Medicine in the Operating Room

A New Age Pioneer Takes Her Intuition
Where Few Have Dared to Go

Cynthia Logan

Unprocessed trauma and withheld anger," says energy healer Julie Motz, "is responsible for 99.9% of all serious illness and disease." Speaking with clarity, confidence, and the quick pace of a native New Yorker, she continues: "Anger is misunderstood; it's not recognized as being the energy that can move us forward."

Motz is the author of the extremely well-written book *Hands of Life*. "This is a must read for any open-minded person suffering from an illness. Julie Motz has captured the essence of mind-body medicine at the cellular level," says Stephen Sinatra, M.D. Motz has received widespread recognition for her groundbreaking and often controversial work as the first non-traditional healer allowed to work side by side with surgeons in operating rooms at such prestigious hospitals as Columbia Presbyterian Medical Center in New York and Stanford University Hospital in California.

Her energy-healing techniques help surgery patients need less anesthesia and recover faster and more completely. She has lectured at Stanford and Dartmouth medical schools, has presented her work at national and international conferences, and has been profiled on *Dateline*, CNN, the *New York Times Magazine*, *New Age Journal*, *USA Today*, and *Ladies Home Journal*.

According to Motz, many of the techniques employed for processing anger (such as hitting a bed with a baseball bat and throwing pillows) are not effective. "Hitting is abusive to your own body and it throws the energy out of the body; what you need to do is move it through the body, so you can access its strength. Anger isn't something you want to get rid of; it's something you want to appreciate and use and have at your disposal."

Motz knows firsthand what it's like to experience the power of anger moving through the body. In processing her own childhood trauma, she

Fig. 25.1. Energy healer extraordinaire Julie Motz.

participated in "Fusion Groups" ("encounters" popular during the 1970s). She credits the work of Mike and Sonja Gilligan as pioneering and inspiring: "Sonja Gilligan's unique contribution was to realize that there are four and only four basic feelings," says Motz. Gilligan correlated them with the four forces in matter (electromagnetism, gravity, nuclear force, and weak force); later, Motz postulated that these "emotional forces" are carried in specific systems, tissues, and fluids in the body.

"Fear corresponds to electromagnetism and is the same as excitement; it is the feeling state of perception and should be carried in the cerebrospinal fluid and the nervous system," she states. "Anger is the emotion of action, corresponding to gravity. If your fear tells you that what you have perceived is safe, your anger carries you forward to get what you desire. If your fear tells you it is dangerous, your anger gives you the energy to fight or flee." Motz says that anger should be carried in the blood and in the muscles.

According to Gilligan and Motz, pain corresponds to the nuclear force, with its power to pull in toward the center. "Pain, in the emotional sense, is self-knowledge," says Motz. "It puts you in touch with the core of your being, and should be carried in the lymphatic fluid and the bones."

Love, the fourth emotional force, corresponds to the weak force, and should be carried in the synovial fluid of the joints and in the bone marrow. The weak force is released deep inside stars like the sun, where they create the energy that keeps us alive. According to Motz, it is this mysterious "weak" force of love that is, in fact, the healing force.

Holding a master's degree in public health, Motz blends Eastern holistic concepts with Western medicine; she employs a number of modalities, none of which she has formally studied for more than five days. Using her intuition, touch therapies, and acupressure meridians, as well as verbal and psychic messages, she helps patients to harness the body's own energy, intelligence, and memory. Such a combination has proved effective for even the riskiest operations, including heart transplants, head trauma, and breast cancer surgery.

Motz, who can feel a patient's emotions in her own body, believes that her ability to identify and interpret physical/emotional signals is something almost anyone can learn. She likens the process to learning how to track with an Indian guide. "At first, you are amazed that he can tell that a deer stood next to this particular tree, or that a fox came through the underbrush at just this point. Then the guide shows you the place where the buds have been stripped from a branch or some dry leaves displaced along the path. It's not that you didn't 'see' these things; they passed across your visual field, just as they did across his. But you didn't 'notice' them, because you had not trained yourself to consider such signs as bearers of important information."

Though it may be something all of us are capable of, Motz's ability to sense emotion within bodily structures is clearly head and shoulders above most of us. Sometimes compared to medical intuitive Carolyn Myss, whom she admires for popularizing the whole concept of energetic healing, Motz points out that she is much less metaphysical than Myss. "I'm much more interested in the physics," she says. Having studied the work of French physicist Louis de Broglie (who posited that all matter emits waves that travel faster than the speed of light), she finds it entirely possible that matter as highly organized as human tissue would not only give off distinctive and identifiable wave patterns, but also be able to receive and identify them. "This is what I believe is going on when I feel another person's emotions in my body," she says.

Motz theorizes that there are two possible ways that enable energy healing to work: "It may be that when I touch a patient, the energy of [his or her] body begins to align itself vibrationally with mine, and the meridian I am touching shifts. The other possible explanation is that when I touch with a loving or a healing intent, that point of contact becomes a magnet for ambient energy in the room, which enters the person's body through my hand or finger at that point. In this theory I am actually attracting a flood of neutrinos that interact with other subatomic particles inside the body to alter the energy flow."

The daughter of a professor of theoretical physics and a director of New York City School Libraries, Motz comes by her intellectual prowess naturally. She is used to vigorous discussion, reads voraciously ("literally anything that's well written"), writes poetry, and lists "thinking about things" as one of her leisure pastimes. Though she considers angels and reincarnation to be "things the frontal cortex creates to soothe us when trauma starts pushing," she also admits having had at least two "interesting" energetic experiences about which she withholds judgment. "I believe the universe is evolving toward order and love, but there's no Deity separate from us, which is actually an Eastern concept."

Her experiences with the Fusion Groups and subsequent work with patients has led her to the wisdom of the heart. "I returned from the world of the intellect to the world of emotions, which bridges the brain to the body," she says. "I reconnected thought and feeling and admitted that the love I had taken in was what had made me smart."

That love has paid off: her services are $250 per hour out of surgery and $175 in the hospital. Well aware of her worth, she has had to prove it within the politics of the medical establishment, and is passionate about both her work and the concept of energy healing within a larger, social context. "Thousands of people believe they can be healed only by having someone saw open their chest and physically touch their heart, as they themselves could never do," she says. "Many other people, like the surgeons I work with, enjoy cutting open chests and moving things around inside. These two groups of people are likely to be coming together for a long time. So long as they do, there is a place for an energy healer."

She further speculates that the high drama and ritual of surgery has deep meaning for both the "rescued" and the "rescuers" in our society, and points out that "[j]ust by coming into the OR, the patient is expressing his love toward the surgeon. Allowing the surgeon to open and enter his body, while he, the patient, is utterly out of control is an even greater act of love." She envisions an operating room where everyone in attendance thanks the patient for being there and for bringing his love and trust into the room to create a healing space for all. "Without him, there would be no focus for our love and our knowledge. Each of us would commit to putting into the space of nonjudgment—in which the surgery would occur—whatever it was we needed to have healed at that time."

Motz believes that brain tumors may originate from "asking the brain to perform a task that is really meant for the body . . . Anger and love, as action and connection, are feelings of the body," she says. "Fear and pain, as perception and understanding, are feelings of the brain. Emotions are the energies that run both our lives and the universe. The more emotional blocks you move out, the more energy you have. It's not having emotions that drains our energy, it's suppressing them. That requires a lot of energy."

Her own exuberant energy is fueled by the macrobiotic diet she has followed for many years, and by what she calls a "habitual reaction to resistance which is, when something is denied me, to press even harder for something even more difficult and out of reach."

Such persistence may have helped save her life; Motz fought bulimia and attempted suicide before she was introduced to the world of feelings through the Gilligans, with whom she formed a film company after earning her master of fine arts in cinematography. At the time, their films featured American history and culture. Today, she muses about buying a video camera and incorporating it into healing.

Meanwhile, her Health's Angels program teaches energy healing to inner-city teenagers, and she is currently working with pregnant teens. "Pregnancy is an opportunity to heal," says Motz, who does not have children herself. She has come to the conclusion that most trauma begins in the womb, and notes that prenatal life is where depression, addiction, and overeating begin.

"The idea that children are a burden is not a truth of the universe; it's something that's passed on from mother to child," she says emphatically. "It's not a sacrifice to carry a child; it's an energizing adventure, a collaboration. What happens during pregnancy is that you relive your own gestation. And, if it was not a happy time for your mother, it won't be a happy time for you."

"The best thing we can do to help our children is to do our own healing," she asserts. She points out that repeated emotional distress and abuse—starting very early on in life and usually ignored or denied by the patient—are key factors in weakening the body and making it vulnerable to chronic disease. "You have to open the wound in order to heal, or any work you do is just going to bounce off the scar."

Since rediscovering and processing buried trauma are not skills widely embraced by our society, Motz suggests simple steps rather than a radical technique such as "rebirthing," which she feels can be unnecessarily violent. "Start by observing things that bother you in life, ones that don't go away, and find out what happens in your body: how does it feel, what age does that feeling come from? Then try to bring up a scene from that age. . . . As people find the courage to confront the things their parents did to them, the more courage they're finding to stand up to the things doctors are doing to them."

While she bemoans the general lack of awareness and sensitivity among many surgeons, anesthesiologists, and attending personnel in the operating room, Motz has great admiration and respect for the dedication and talent of most and the brilliant genius of a few. She observes that different surgeons doing the same operation are like "different musicians playing the same piece; they start and finish at the same place, but that's all."

Motz sees the trend toward complementary healing and self-care increasing: "A year ago I drove across the country and stayed in small bed-and-breakfasts; I was amazed at the people using herbal remedies and acupuncture; there seems to be a general interest in alternative therapies." As Dr. Christiane Northrup asserts, Motz feels that medicine is largely a parental practice and notes that "the patient and the medical team create a mutual experience that allows everyone involved to heal emotional traumas. The importance of a person's emotional history to his physical well-being has long been acknowledged in medical circles, but once that person passes through the doors of the medical institution, it is largely ignored."

What is the investment in continuing to act as if feelings have so little impact on the internal actions of the body? To help educate patients and professionals, Motz plans to conduct both an East Coast and a West Coast training in the near future. "I'm interested in creating healing communities and I'd like to have people on both ends of the country that can discuss and develop this work," she says. Her book succinctly summarizes the mind-body split that so many people feel modern medicine has helped to create: "Perhaps our technological failure to deal adequately with chronic disease on every level of society may be a great gift. In our desire to end this violence to our bodies, we may at last extend to each other the love that gives us the courage to feel and to know ourselves completely. This is what it means to put ourselves in the hands of life."

26 Getting Left and Right Brains Together

Author and Doctor Leonard Shlain Believes the Future Is Where Art and Physics Intersect

Cynthia Logan

Negotiating the freeway toward San Francisco's California-Pacific Medical Center, where he is chief of laparoscopic surgery, Dr. Leonard Shlain presses the eject button on his new car's console (past models were Jaguars, but now that he's a conscious consumer, it's a Prius). Out comes *The Count of Monte Cristo,* a story that has had him riveted ("I actually had to pull over at one point to find out what happened," he admits). But he's going to be operating on someone's carotid artery this morning and needs to concentrate; so, for the moment, the story is on hold. Is his patient left- or right-handed? Was the injury to the left side or the right side of the brain? "This is an important factor in how I'll approach my job," he explains.

"Carotid procedures hold a special fascination for me; operating on carotid arteries requires that I understand how the brain works."

Shlain has long been fascinated by right/left hemisphere differences in the brain, as well as by modern art and science. Ruminations concerning the puzzle of consciousness, the right/left split and the connections between cubism and relativity "tumbled like clothes in a dryer" through his mind, and he's been ironing out the wrinkles over the years. His thoughts emerged as

Fig. 26.1. Surgeon and humanist Dr. Leonard Shlain.

183

an award-winning book in 1991. *Art and Physics: Parallel Visions in Space, Time, and Light* (HarperCollins) proposes that innovations in art prefigure major discoveries in physics; it is presently used as a textbook in many art schools and universities. The book covers the classical, medieval, Renaissance, and modern eras. In each, Shlain juxtaposes works of famous artists alongside the paradigm-shifting ideas of great thinkers. Pairings include Giotto and Galileo, Da Vinci and Newton, Picasso and Einstein, Duchamp and Bohr, Matisse and Heisenberg, and Monet and Minkowski.

The youngest of four children of first-generation Russian emigrants, Shlain loved building model airplanes and enjoyed drawing as a child, and fancied himself a budding artist. He considered psychiatry as a career, but chose the dramatic life of a surgeon (think *Magnificent Obsession*): "It was romantic, challenging, and intensely exciting."

Also an associate professor of surgery at UCSF, Shlain knows first-hand what it's like to be under the knife. At thirty-seven, he had passed his Boards, become a fellow of the American College of Surgeons, and had a university teaching position, a wife, three children, and a "tiger-by-the tail" burgeoning practice. Things were going according to plan when he found himself "sitting on the edge of a hospital bed dressed in the half gown of a post-surgical patient"—he'd just been told that a biopsy had come back malignant for a non-Hodgkin's lymphoma.

His subsequent treatment and recovery led Shlain to participate in a "death and dying" seminar. "An organizer of one of the workshops was familiar with my recent encounter with the grim reaper and thought it would be neat to have a surgeon provide his perspective from both sides of the scalpel," he recounts. His story was later included as a chapter in *Stress and Survival: The Realities of a Serious Illness* (a compilation edited by Charlie Garfield). Unbeknownst to him, Stanford's radiation department had copied his chapter to hand out to incoming patients, and a medical school had made it required reading for junior students beginning their oncology rotation. "My career as a writer emerged from the single worst experience of my life," notes Shlain. After a fortuitous encounter with a New York agent, he learned that eight major publishers were interested in that single chapter becoming a book.

"For the next year I was a man possessed. I wrote early in the morning before surgery, on vacations, on weekends, and while waiting for cases to begin in the surgery suite." Shlain says he approached the art of writing as he had approached the acquisition of the skills necessary to become a surgeon.

Was the Great Pyramid of ancient Egypt intended as nothing more than a huge mausoleum for the Fourth Dynasty pharaoh Khufu (Cheops) in 2550 BCE? Or could it have been something more interesting, such as a temple of initiation or a power plant?

This small ivory statue is thought to be a representation of Egypt's Pharaoh Khufu (2550 BCE), who is credited in orthodox Egyptology with directing the construction of the Great Pyramid at Giza. (Photograph by Greg Hedgecock)

This Mayan pyramid at Palenque, in the Yucatán (left), in many ways resembles the Third Dynasty pyramid built by Pharaoh Zoser at Saqqara in Egypt (right). Both are step pyramids and both contain subterranean chambers with descending corridors. Could this resemblance suggest a common mother culture?

Cultural comparisons between the ancient Near East and pre-Columbian America suggest that a vanished, mid-ocean source was common to both cultures. Was that ancient source Atlantis? (Painting by Tom Miller)

The 204-foot-high stone step pyramid at Saqqara in Egypt, generally believed to have been built in approximately 2800 BCE, overlooked the great ancient city of Memphis.

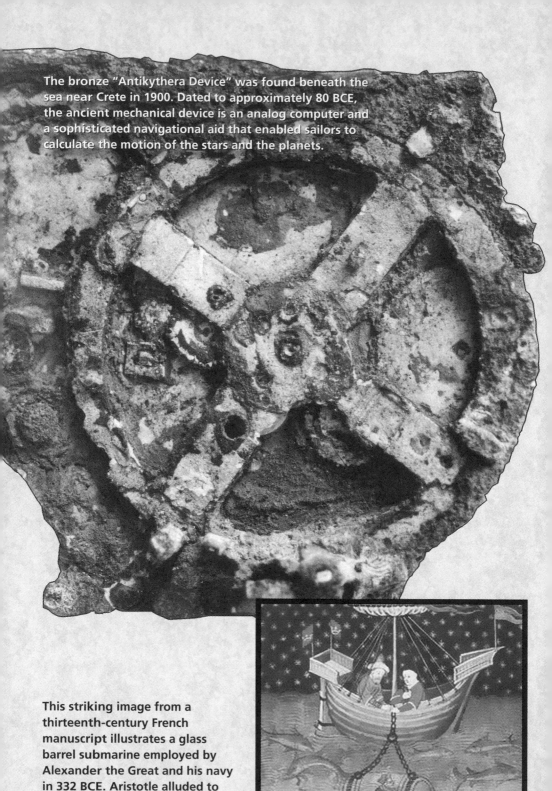

The bronze "Antikythera Device" was found beneath the sea near Crete in 1900. Dated to approximately 80 BCE, the ancient mechanical device is an analog computer and a sophisticated navigational aid that enabled sailors to calculate the motion of the stars and the planets.

This striking image from a thirteenth-century French manuscript illustrates a glass barrel submarine employed by Alexander the Great and his navy in 332 BCE. Aristotle alluded to these vessels also, calling them "submersible chambers."

This "Baghdad Battery" was discovered at Khujut Rabu, just outside Baghdad, and dated to 250 BCE, more than two thousand years before the officially decreed invention of the electric battery by Alesandro Volta in the early nineteenth century. It is believed that, among its other functions, it may have been used to electroplate items with gold.

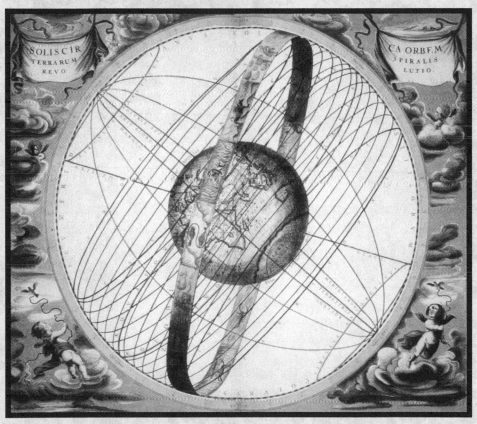

This Renaissance depiction of the earth with the zodiac shows the regions traced out in the heavens during the precession of the equinoxes. (Artwork from the seventeenth-century "Atlas Coelestis")

Throughout the ages, those who challenge the prevailing status quo have often been persecuted, whether it be Galileo in the Inquisition of the 1600s (shown here) or Immanuel Velikovsky in his trial by academia in the twentieth century. (Painting by Cristiano Banti, 1857)

Did Newton truly understand how his apple would have behaved in orbit?

These ancient Dogon villages are located along the Cliff of Bandiagara in the Sangha region of Mali, Africa. (Photograph by Galen R. Frysinger)

A NASA artist's representation of our galaxy, the Milky Way.

Above: Originally over twelve feet high, Megalith X-1 is the largest stone megalith at Nabta Playa in Egypt. Like many such once standing stones at Nabta, it was broken or perhaps intentionally cut apart eons ago. The immense Nabta Playa structure, it seems, was part of an ancient star-viewing platform. (Photograph by Thomas Brophy)

Right: Worked and finely shaped megaliths from the top of the complex known as "Structure A" at Nabta Playa, Egypt. (Photograph by Thomas Brophy)

Below: A quartzite sandstone monolith at Nabta Playa.

Artist Tom Miller's conjecture of what an ancient atom blast on the Asian subcontinent may have looked like.

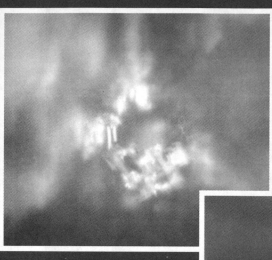

Left: Distilled water microscopically photographed at the point of ice formation.

Below: The same water crystal (displayed in the color plate at left) after being exposed to a picture of a child.

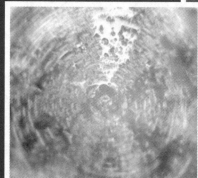

Above: Contrast this pattern that a water crystal forms after being exposed to rock music with that of the photo to the right.

Right: Exposure to Bach's "Air for G String" forms a very different and very beautiful water crystal.

(Photographs from *Messages from Water* by Dr. Masaru Emoto)

Playing with the Levitron, an antigravity top. It achieves sustained levitation through a combination of magnetic repulsion and centrifugal force.

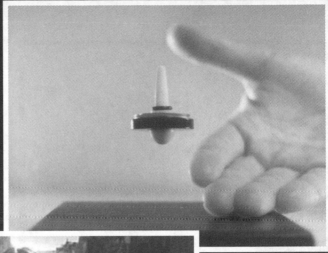

Above: Susumi Tachi's invisibility cloak, which attempts to camouflage its wearer by electronic means.

Right: When turned by light energy alone, the black and white paddle wheels of this radiometer can be stopped with channeled mental energy, proving that mind can influence matter.

Artist Tom Miller's conjecture of an antigravity future.

An artist's conjecture of Tesla's X-ray machine.

Sonofusion researcher Rusi Taleyarkhan, of the Oak Ridge
National Laboratory, Tennessee. (Photograph courtesy of Oak
Ridge National Laboratory)

An artist's conception of an invisible fighter jet. (Illustration by Randy Haragan for *Atlantis Rising*)

Near Gakona, Alaska, HAARP antennae stand 65 feet high and are spaced 80 feet apart in eight columns of six rows. Ostensibly these antennae are used to research the upper ionosphere but could they be, instead, tools used to manipulate global weather patterns? (Photograph courtesy of U.S. Air Force)

The clouds of Venus.
(Photograph courtesy of NASA)

Hero astronaut John
Glenn enters *Friendship 7*.
(Photograph courtesy of NASA)

The Apollo 14 mission takes Edgar Mitchell on a voyage of immense personal discovery. (Painting by Tom Miller)

"I knew that proficiency begins with considerable practice and the emulation of experts. I had always been a voracious reader [he particularly likes Dostoevsky, and enjoys Hemingway, Steinbeck, Melville, and Dickens]; even when I was in the midst of the most demanding rotations of my surgical training, I always had a paperback in my back pocket," recalls Shlain.

Seven years after the success of his first book, a second emerged: *The Alphabet Versus the Goddess: The Conflict Between Word and Image* (Viking) hit the national bestseller list within weeks after publication in 1998. Conceived during a tour he took to Mediterranean archaeological sites in 1991, the book discusses his theory that with the development of the alphabet and the rise of literacy, a right-handed, left-brained dominance was created (the hemispheres of the brain control opposite sides of the body), birthing a patriarchal culture that diminished women, goddesses, and sacred images. While he appreciates the gifts of literacy and literature, Shlain points out that the writings of Marx, Aristotle, Luther, Calvin, Confucius, and many other male writers advanced and perpetuated patriarchal values, particularly through the Bible, the Koran, and the Torah. "In the ancient world, during polytheism, people weren't killed for their religious beliefs," he says. "With the rise of Christianity, Islam, and Judaism, we began fighting over whose One God was The Right God."

In a fascinating discussion enhanced by anatomical understanding, Shlain postulates a difference between the eye and the ear, between speaking and listening: he characterizes the eye's function and speaking as being more active and male, whereas he sees the ear's architecture and listening as being more receptive, or female. As a society, we focus more on speaking and seeing than we do on listening and hearing. The human eye contains specialized "cone" and "rod" cells, which Shlain associates with left and right sides of the brain, respectively. Interestingly, men's eyes have more cone cells (which home in on close focus), while women's contain more rod cells, which see images as a whole gestalt. Thus, the sexes literally see things differently. (Finally, an explanation for why men often cannot find something in the refrigerator; as for why women talk so much—they see more, so there's more to say!)

A passionate person, Shlain enjoyed a tumultuous courtship with his wife of seventeen years and endured an equally tempestuous divorce. Single for seventeen years afterward, he remained fascinated by the differences between men and women and, though reticent to "get back in the game," always thought most people unconsciously seek soul mates. He met his second wife, Ina, a judge, on a blind date and, for him, the myth of a divided soul finding its other half has become reality. "We're two control freaks who don't like conflict," he says, adding that "somehow, that really works for us. When we cook together, I, in the manner of my professional training, place my hand palm-up without looking away from the slicing and dicing and bark, 'Tomato!' Ina laughingly intones: 'Motion overruled.'"

Every chapter in each of Shlain's books is titled with provocative, often juxtaposing dyads, such as Athens/Sparta, Illiteracy/Celibacy, Menarche/Moustaches. "It goes with the right/left duality that so characterizes our reality," he says. His latest work, *Sex, Time, and Power: How Women's Sexuality Shaped Human Evolution*, explores why *Homo sapiens* evolved so differently from other animals. His conclusions—that women evolved the power of choice regarding whether or not to engage in sexual intercourse once they connected it with childbirth, that menses evolved to teach the species how to tell time, and that women's loss of iron through menstruation and other avenues resulted in a balance of power between the sexes (as men learned to hunt and bring home the bacon that restored her depleted reserves)—are well researched, elegantly stated, and just as plausible as other explanations.

Critics have focused their complaints about Shlain's ideas on the fact that someone other than an artist, scientist, anthropologist, paleontologist, or linguist has written with such authority on subjects that they believe should be pioneered by academic experts in those fields. "Important contributions are made to many areas by laypersons." suggests Shlain. "Theories are often put out there as 'Just So' stories and are later proven to be valid." Besides, as he writes in the introduction to *Sex, Time, and Power,* as a surgeon he has had a lot of time to deeply ponder the mystery of blood and the fact that, though the variables of most of twenty-six numbers on a chemistry panel are gender neutral, men normally have a 15 percent higher concentration of circulating red blood cells than a healthy woman has. The understanding of hemoglobin and the role that iron plays in the dance between the sexes became a quest that Shlain pursued meticulously.

Recently popularized theories about the Holy Grail having been Mary Magdalene herself point to the reemergence of the "Divine Feminine" or, as some prefer, the return of the Goddess. Shlain's Mediterranean archaeological tour had ended at Ephesus, where the ruins of the largest shrine to a female deity in the Western world, the Temple of Artemis, lie. "Our guide had told the legend of Jesus' mother, Mary, coming to Ephesus to die, and pointed out the hillside on which her remains were purported to have been buried," he writes. Asking himself why, if the legend was true,

Mary would have chosen a place sacred to a "pagan" goddess as her final resting place, he began to question what had caused the disappearance of goddesses from the ancient Western world. "There is overwhelming archaeological and historical evidence that during a long period of prehistory and early history both men and women worshipped goddesses—what in culture would have changed to cause leaders in all Western religions to condemn goddess worship?

The Alphabet Versus the Goddess isn't all "alphabet bashing." With unprecedented advances in image technology, Shlain (who is himself right-handed but ambidextrously talented) thinks we are witnessing the dawn of a new language, one that is reintroducing the voice of the right hemisphere. "I hear a lot of talk about how we're destroying life as we know it. How about a different scenario—that we are in the process of evolving into something different?" he asks. In his mind, the rise of images through the media, along with the use of computers, is a positive evolutionary development. Where writing has traditionally required the use of the right hand and the left brain, keyboards and screens engage both, like a musical instrument. (Typing was a step toward hemispheric coordination, but was predominately engaged in by female secretaries.) Computer keyboards are tapped by both sexes. "Just think of all those male left hands on keyboards, stimulating their right brains," enthuses Shlain, who also feels that relationships are reinforced by technologies like e-mail, cell phones, and beepers.

So, what's the doctor's prognosis for the world? "We have a whole new set of problems now," he tells audiences throughout the United States and Europe. A keynote speaker for such diverse groups as the Smithsonian, Harvard University, the Salk Institute, Los Alamos National Laboratory, NASA, and the European Union's Ministers of Culture, Shlain's message about the future is upbeat. Besides the merging of the hemispheres (a global metaphor, perhaps?), Shlain points to historical shifts that he likens to Hegel's philosophical "Thesis, Antithesis and Synthesis" model. Just as the Renaissance gave way to the Reformation, which then transformed into the Enlightenment, so the 1960s were completely converted in the 1990s, which appear to be headed in a completely different direction in the 2000s. "The flood of images in which we now live is restoring a long-lost balance between our linear left-brain hemispheres and the visual right ones, bringing an end to a five-thousand-year reign of misogyny."

Himself a Renaissance man, Shlain's next book concerns Leonardo da Vinci, but it will be "nothing like what's on the market now." In addition to having raised three highly creative and contributory children (of his many roles in life, he most cherishes that of father) and having won several literary awards for his visionary work, Shlain also holds a number of patents on innovative surgical devices having to do with stapling, cutting, and cauterizing. Like his scalpel, Dr. Leonard Shlain uses words and concepts to fulfill the mandate he has hanging above his writing desk: a quote from Franz Kafka urges writers to create books that can be wielded like a pickax to shatter the frozen sea within the reader's mind. "If a book didn't change the way the reader thought about the world, then Kafka deemed it not worth writing," he reports. Using this as his credo, the successful author finds unique metaphors for conveying complex concepts and weaving in a dizzying array of facts to support them. In doing so, he accomplishes his stated mission to "set the reader's mental ice floes grinding against each other."

Whatever one's take on Shlain's theories, reading his books is a provocative and thoroughly enjoyable adventure. Describing himself as a storyteller by nature, the man loves the "luxuriant diversity" of English. He has painstakingly avoided technical jargon, yet throws in the occasional unfamiliar noun, verb, or adjective. "At times, I could not restrain myself from trying to rescue a few of my favorite words from what I fear may be their impending extinction." Get out your dictionary, sit back, and prepare to be both educated and entertained.

27 X-Ray Vision and Far Beyond

Troy Hurtubise's Amazing Lights Are Said to See through Walls and to Heal as Well

John Kettler

Superman had it, as did Ray Milland in Roger Corman's *The Man with X-Ray Eyes,* along with other assorted superheroes, and it has been rumored that even Nikola Tesla had a machine that could accomplish it. So-called X-ray vision, or the ability to see through walls, however, has remained pretty much a science-fiction fantasy—unless you count the kind hospitals and your dentist can do. All that may be about to change, though, thanks to the efforts of a maverick Canadian who has apparently invented a machine that can not only actually see through brick walls but do many other amazing things as well, maybe even cure cancer.

Thomas Edison once characterized the invention process as being "1 percent inspiration and 99 percent perspiration." Sometimes, though, the inventor catches a break and literally dreams up the invention, complete in every detail. The challenge then becomes externalizing that clearly perceived internal blueprint. What if, however, you had zero expertise in the exotic technical fields covered by your invention? Further suppose that the device you're building goes beyond even the boundaries of cutting-edge science. Now what?

Troy Hurtubise, forty-one, is a prolific inventor from Ontario, Canada, whom some readers may know as the designer of the Ursus Mark VII suit, body armor strong enough to withstand grizzly bear attack, plus the revolutionary Fire Paste thermal protection system (demoed on camera by Troy holding a ½-inch thick tile atop his head while applying

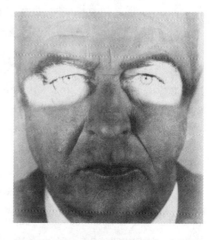

Fig. 27.1. Ray Milland in The Man with the X-ray Eyes.

Fig. 27.2. Inventor Troy Hurtubise.

a 3,600-degree Fahrenheit torch flame to the other side—for ten minutes!) and the astounding Light Infantry Military Blast Cushion (LIMBC) system. With seven cameras watching, a foot-square by four-inch-thick LIMBC sample mounted on a vehicle door stopped some forty highpowered rifle (.370) shots cold—ceramic armor might've taken two hits—then survived a direct RPG (rocket-propelled grenade) hit with only a minor dent in the door. The non-LIMBC door was first sieved by the rifle fire, then blown apart by the RPG. These and other of Hurtubise's creations have been repeatedly featured on the Discovery Channel and The Learning Channel programs, both in the States and in Canada.

This tough ("I was shot twice, stabbed six times" during years spent in the mountains), outspoken man—whose natural resources technology degree (bear behavior specialist) from Sandford Fleming College, Lindsay, Ontario, Canada, is the equivalent of a bachelor's here—set out to invent a way to find people in collapsed buildings, a wholly laudable thing, only to unexpectedly detour into the black realms of anti-stealth, spooks, and malefic energies. That was the beginning of his ongoing odyssey, one so bizarre that even Hollywood would reject it as a script.

In an exclusive series of blockbuster articles, Phil Novak of North Bay, Ontario, Internet news site www.baytoday.ca, dropped a consciousness bomb that even now is rocking the world—not in the mass media. Rather, the action is whirling and churning behind the scenes: on the Internet, in governments, militaries, intelligence agencies, scientific labs, and corporate boardrooms, all caused by one man driven to invent while trying to keep his family housed and fed.

SEEING THROUGH WALLS

Disaster search and rescue becomes both easier and far more lifesaving when victims can be located quickly, extracted, and treated for their injuries. Progress in this area has been considerable, first with specially trained dogs, remote cameras, and microphones, now with small robots. If certain experiments are credible, we may soon have remote-controlled search rats, too. The technical approach favored by top sensor firms, though, is to use technology to "see through" walls, to which end much work has gone into developing microwave (à la the archaeologist's ground-penetrating radar) and ultra-wideband devices to detect things like human heartbeat and form three-dimensional images of what lies beneath building rubble. Hurtubise thought there might be a better way, in a different part of the spectrum. His used light—lots of it!

THE ANGEL LIGHT

The Sunday, January 16, 2005, headline said it all: "Hurtubise says invention sees through walls—BayToday.ca exclusive" (www.baytoday. ca/content/news/details.asp?c=6657). As if this weren't enough, readers were next informed that the invention reportedly "defied all known laws of physics," and two paragraphs later we're told "the device detects stealth technology." Whew!

The device in question would make any mad scientist proud or prop house jealous, but witnesses—reportedly including French government representatives, the former head of Saudi counterintelligence, and the inventor himself—say this one works. The "how" is largely unknown, the result of the incredibly advanced technology on the one hand and Troy Hurtubise's judicious silence on the other. What is known, though, is that the "Angel Light" uses three discrete energy forms—light, plasma, and microwaves—in some heretofore unseen combination (a Ph.D. physicist says Hurtubise "fused light") to reportedly perform the impossible.

After Hurtubise had the same dream for weeks, then spent some eight hundred to nine hundred hours and tens of thousands of dollars on exotic components and power supplies, the Angel Light was born. In tests with it, as reported by Phil Novak and the inventor, Hurtubise "saw" right through the garage wall, and was able to not only read the license plate on his wife's car but also see the road salt encrusted upon it. Great, right? Not quite. You see, the device had lethal biological side effects. Curiosity almost killed the inventor, and definitely reduced the goldfish population. How? He put his hand briefly in the beam, an ill-advised action, the consequences of which were that, in his own words, "I lost use of my hand for a year, plus puking blood, losing weight, and my hair started to fall out." Grim! Worse was to come, though, for when scientific buddies at

MIT advised him to put a small goldfish tank in the Angel Light's beam for biological testing, they blithely assumed he'd be behind some sort of protective wall. He wasn't, so got "splashed" by the beam (partial reflection from the tank) on several occasions. It was much worse for the goldfish; thirty goldfish went belly-up in minutes after being exposed to the beam for a mere three millionths of a second. This scenario is powerfully indicative of acute radiation exposure and became a source of great personal and moral concern to the inventor.

ANTI-STEALTH

Many people believe that stealth is the crown jewel of American military technology, an unbeatable means of achieving and maintaining battlefield dominance. In reality, the technology is beatable and has vulnerabilities. Troy Hurtubise, aided by a network of friends, apparently found another one in the Angel Light. Using a borrowed sample body panel from the canceled RAH-66 Comanche stealth recon/attack helicopter and a police radar gun, he found that his light would negate stealth—with yet another stunning side effect.

DIRECTED-ENERGY SUPERWEAPON

The side effect? Fried electronics! In the above test, the Comanche panel was mounted on a radio-controlled toy car, running on a track located on what we could call a Native American reservation. Not only was stealth negated, but the beam killed the car stone dead as well.

In later tests, all sorts of expensive electronics were burnt out by the beam. At a stroke, Troy Hurtubise had a fully operational non-nuclear electromagnetic pulse weapon! It was this potential that brought the French to his door. In a lengthy telephone interview with the writer, whose former professional background includes work on such weapons, the inventor said that it was obvious to him that the French wanted the Angel Light as a truly effective strategic defensive system (contrasted by him with the ludicrous, in his view, U.S. missile-based approach), to which end they provided him some $40,000 (Canadian), specifically to increase the ceiling from some 70,000 feet clear up to low earth orbit. That part was doable, but the effort foundered when Troy Hurtubise, try as he might after months, couldn't figure out how to get rid of the deadly biological side effects.

While perfectly willing to sell the French (the United States was not interested) a strategic defense weapon (could've made millions), he wasn't about to let any country or group get its hands on a literal death ray, which is precisely what the Angel Light was as long as the biological side effects remained. The inventor described a hypothetical scenario in which an entire division was zapped by the Angel Light, to the utter confoundment of the victims, who'd feel nothing, only to be devastated an hour later as their bodies dissolved from within. Not that it would matter to them, but every electronic device would be rendered useless at the time of the strike, too. Faced with an irresolvable moral problem and not money-focused (he told the writer he wanted only "a house, a pickup truck, and a proper lab"), he ceased all experiments with it and dismantled the Angel Light. That might've been the end of the story, but greater things awaited.

THE GOD LIGHT . . . SIDETRACKED

He didn't know it then, but Hurtubise was about to be diverted—massively—from his planned rescue device and into realms unimaginable to him. In the process, he'd lose the Angel Light's ability to see through walls, but would also no longer have a death ray. What he got in exchange was to many beyond price: a true healing-life-enhancing machine! Phil Novak's

Fig. 27.3. Troy Hurtubise's amazing lights are said to see through walls and to heal as well. (Photograph courtesy of BayToday.ca)

next stories chronicled the sea change: "Angel Light ascends to God Light. Parts One and Two. Bay Today.ca exclusive," Wednesday, May 11, 2005, and Thursday, May 12, 2005 (www.baytoday.ca/content/news/details. asp?c=8267) (also 8271).

UNEXPECTED HELP/EVEN MORE UNEXPECTED DIRECTION

Ironically, for an inventor who by his own admission "can't program a VCR" and "doesn't understand computers," his deliverance came from Germany via a Web cam (presumably, a friend's), in the form of help from a German physicist, an electronics engineer, and an electrician. This became necessary after his would-be helpers realized that he couldn't read the schematics they sent him. And it was through this peculiar collaboration that he found himself suddenly thrust into oncology, after being referred to a Toronto cancer researcher, who brought over lab mice (with known tumors) for testing under the "God Light." The results were, in Hurtubise's words, so "staggeringly positive" that the inventor has thrown his lab open to "anybody of scientific credibility in the scientific world, who works on, say, Parkinson's, AIDS, Alzheimer's, or MS." How positive? Specimen C-12, following a God Light exposure of twenty minutes and seven seconds, experienced a "27 percent reduction in the tumor." C-12 had previously undergone radiation therapy. More tellingly, Specimen H-27, with a brain tumor, no prior radiation therapy, and a God Light exposure of eighteen minutes and thirty-three seconds, experienced a "12 percent reduction," and in both cases, the cancer's progress was completely arrested, without adverse side effects, in a fifty-six-hour observation period.

Savvy readers will no doubt rapidly make the connection between the God Light and the earlier work of Priore, which relied on a combination of microwaves and strong magnetics, as detailed by Tom Bearden at www .cheniere.org. Asked about this by the writer, the inventor concurred that there are indeed major parallels, despite the spectral region difference, but was at pains to point out that he was also drawing upon a vast, previously scattered body of reputable work done worldwide on light at various frequencies in therapeutic applications. The equipment list supplied to Phil Novak by Troy Hurtubise confirms that this is indeed a light-based therapeutic approach, evidently in conjunction with magnetics, plasma, sound, and what may be microwave energy.

PERSONAL HEALING

Having seen firsthand what the God Light did for the mice, the inventor, despite a very bad prior experience with the Angel Light, couldn't resist making himself a human guinea pig again. This time, despite a burning sensation the physicist later opined might be related to cell regeneration, Troy Hurtubise hit the jackpot, being rapidly healed of his hair and weight loss, his hands (also affected by arthritis from pounding out the Ursus suit's armor plates) were healed, and his energy was restored, too. Just as well, for this driven man is in the lab a whopping twenty-one hours a day.

Some readers may have seen the famous Kirlian photograph of a cut leaf (showing the outline of the complete leaf over an hour after the top of the actual leaf was removed), but Hurtubise trumped that. How? By regrowing a flower on the freshly decapitated stem of a potted plant! Time to begin flower regrowth? Three hours! As if that weren't enough, he put seeds from the difficult-to-germinate (three months) Colorado blue spruce under the God Light—and got seed germination in a week. Phil Novak said in the interview that he had personally daily observed the spruce seed germination experiment over a period of two weeks and was astounded.

HEALING OTHERS

Internet radio-show interviews spread the news of the God Light's amazing healing properties fast, and in no time the inventor was besieged by requests from the afflicted, including a local man, Gary, who suffered from Parkinson's. Fearing all sorts of repercussions (including total ineffectiveness against the disease) but morally unable to say no, Troy Hurtubise let the thirty-seven-year-old come over for free treatment. Two hours and forty minutes of total treatments later, the man skipped up the driveway, rejoicing "I am a man of twenty again."

Such a flagrant breach of human testing protocols greatly disturbed the physicist, but the inventor's focus is on healing and saving lives, and when his sister-in-law came to him with breast cysts (from fibrocystic disease) and begged for treatment in hopes of avoiding yet another round of scarring surgery to remove so-far benign cysts, he had to help. Phil Novak and his photographer were invited guests when her right breast underwent a single ten-minute treatment, after which she reported, "It tingles. Something is definitely going on." Her brother-in-law's guaranteed "reduction in forty-eight hours" wasn't wrong either. When Phil Novak spoke to her forty-eight hours after the God Light treatment, she reported the two cysts had each shrunk from the size of a quarter to the size of a nickel; by the weekend, they were gone altogether. Side effects? Brief, mild nausea. In a phone interview, the reporter said he "was under the God Light myself" on "two different occasions" of "five minutes each." Target was a "lipoma on chest. Was gone two days later."

HARASSMENT AND WORSE

Troy Hurtubise brims with self-confidence, is physically strong from all those years spent in the mountains working with bears, and is morally tough in his utter determination to do what he deems right. Lesser men would've broken long ago under the incredible strain. Why? His lab's been bugged, his phone tapped, people have repeatedly tried to steal his inventions (thwarted by technical sophistication beyond lab analysis), and he and his family have been repeatedly threatened with death. Part of this seems to be tied to conflicting spook issues arising from the diverse and often influential nature of his foreign visitors. With his lab open practically around the clock, and with so much to do, the inventor doesn't have the time or energy to vet visitors, many of whom drop in unannounced. His security's tight, though.

28 The Biology of Transcendence

Do Newly Discovered Retrotransposons Hold the Key to Our Liberation?

John Kettler

That humans have much more intelligence than we can effectively use has long been clear (a fact that our behavior should make obvious). Experts calculate that even an intellectual giant like Albert Einstein used only 5 percent of his theoretical mental potential. When it comes to the three million base pair human genome, the situation is even more extreme, claims scientist Colm A. Kelleher, Ph.D., National Institute of Discovery Science.

In his paper "Retrotransposons as Engines of Human Bodily Transformation,"* Kelleher argues, "Only 3 percent of human DNA encodes the physical body." In other words, 97 percent of three million base pairs apparently do nothing whatsoever. Dr. Kelleher argues that they simply await activation by retrotransposons—genetic structures that may be simply thought of as jumping DNA. Not only does he cite a confirmed case of retrotransposon activation of previously unused DNA, but he also argues that this may well underlie a number of observed metaphysical and religious phenomena, to include age reversal, levitation, transfiguration, and maybe even ascension. His paper proposes a number of laboratory tests to prove the relative commonness of DNA activation via retrotransposons.

Needless to say, the acute underuse of the human genome is driving some scientists mad, never mind Dr. Kelleher's notions. They rail, "Nature is not profligate!" What this means is that nature never gives more than is needed for an organism to function in its environment. What, then, are we to make of this surplus of potential?

The secular humanists and rational materialists describe and treat our species as a thing, as a biomechanical and biological machine, albeit one of great complexity and marvelous design, but what if they're wrong? What if humankind, as understood by them, is not so much an incrementally improving variant of a well-established, time-tested design, but is in fact a kind of

*Journal of Scientific Exploration 123, no. 1 (Spring 1999): 9–24.

Fig. 28.1. Research scientist, biochemist, and avid researcher of scientific anomalies Dr. Colm A. Kelleher.

seething mass of largely unrealized, long deliberately suppressed higher potential? Could this be true? As it turns out, there exists what is, at the very least, a remarkable array of circumstantial evidence pointing toward such a conclusion.

MAN HELD BOUND

If we believe ancient language expert and researcher Zecharia Sitchin, humans were created as a slave race for the Annunaki "gods" (ETs) and were conceived from the union of their carefully detuned DNA and that of the most advanced life-form on earth, a protohominid. The resultant being was systematically hobbled in terms of developing its true potential, but that plan was wrecked when one of the Annunaki decided the new life-form deserved a better fate and tampered with the originally planned genetics. Human beings proved to be anything but docile and caused the Annunaki no end of grief, leading to a decision to let them perish in global floods caused by huge weather upheavals tied to the approach near earth of the Annunaki home planet Nibiru in its highly elliptical orbit. This dastardly scheme was thwarted when the same "god" spilled the beans to one of his favored humans, as detailed in the ancient text *The Epic of Gilgamesh,* directing him to create a vessel for surviving the flood. We know it, in a later form, as the familiar Bible story of Noah's ark.

According to this line of thinking, in Genesis we see a different account of human creation, this time one of initial perfection, in which Adam and Eve walked and talked with God and wanted for nothing, save the forbidden fruit from the Tree of Knowledge, something that, by the way, is a recurring artistic element in ancient Near Eastern art and sculpture. According to the texts, it was by partaking of this fruit that one became or stayed a god. But they ate the fruit—and paid dearly. "By the sweat of thy brow thou shalt earn thy bread" and "In pain and misery shalt thy children be born."

Throughout history and even earlier records—retained as tribal memory, myth, and legend—one of the dominant themes has been the ongoing war between the forces of harmony, advancement, and upliftment and the forces of chaos, destruction, and all-consuming control of everyone and everything, with Hollywood's major recent reenactment being Peter Jackson's *Lord of the Rings* film trilogy, followed by George Lucas's *Star Wars* installment "Revenge of the Sith."

Attorney William Bramley (a pseudonym), in his meticulously documented and carefully argued book *The Gods of Eden,* attempts to make the case that this has been so since the dawn of history and continues to this day. One of his more controversial arguments has to do with the rise of the familiar figure of Death, scythe in hand. Bramley makes the startling suggestion that this figure probably arose as a result of large-scale negative ET biowarfare intervention in Europe at the time of the Black Plague, noting numerous period woodcuts and descriptions of flying objects in the skies, and he goes on to observe that where the accounts said the objects were seen, the plague broke out. Some apparently even indicate the presence of figures on the ground in the fields, with sounds heard from them likened to the *swish* of a scythe through grass. Bramley, not unreasonably, suggests that this could be troops in biohazard suits using hand sprayers. A shambling figure, scything sounds, followed by mass death? Hmm. Whether or not you buy his argument, what is unavoidably clear is that the Black Plague devastated Europe, killing some 33 to 50 percent of the population where it hit—effectively handicapping the growth of civilization there for a long time afterward.

If ancient languages and old woodcuts, though, aren't enough, then some may want to consider what an alleged ET contactee Laura Knight-Jadczyk was reportedly told by the Cassiopaeans (www.cassiopaea.org/cass/supernovae.htm) about a brutal intervention into the genetic destiny of man by the Lizard ET race (not unlike that described by Valdemar Valerian in *Matrix II,* David Icke, and others). According to this scenario, humans still bear the marks of that intervention, in the form of those two lumps (occipital ridge) at the base of our skulls. Knight-Jadczyk says this is the so-called mark of Cain.

Evidently, according to this line of thinking, the Lizard-orchestrated restriction of man's DNA was nowhere nearly sufficient to provide the required levels of control, yet still, everywhere we look we find new controls and restrictions, while we are systematically denied real information, real knowledge, and real power over our own destinies. This particularly applies to the notion that we have any sort of existence beyond three-dimensional space or that we can opt out of a carefully constructed, ruthlessly maintained artificial reality like that presented in the *Matrix* movies.

The seemingly overwhelming grip that the Powers That Be have on nearly everything and everyone is daunting enough to contemplate, but that, according to some, is merely the tip of the iceberg. In myriad ways unimagined by most, that control is both cemented and reinforced. *Matrix III,* by Valdemar Valerian of Leading Edge Research (www.trufax.org), devotes over a thousand pages to detailing a host of means by which we are controlled, ranging from additives in our food, water, and air all the way through sophisticated, individually targeted electronic mind control and beyond.

BIG CHANGES ALREADY

Ironically, however, even as serpentine coils of control seem to wind ever tighter around us, the "snake" has apparently suffered a grievous self-inflicted wound. The Internet (created as a means of reliable, survivable communication even after a nuclear strike), it appears, may have emerged as the greatest breakthrough in uncensored global information exchange ever seen—at a stroke nullifying a control structure painstakingly constructed over hundreds of years. Now, almost anyone can voice his or her views to people almost everywhere—without the say-so of reporters, editors, or publishers, without the access to the media or the money otherwise needed to get into print, without the pundits, the opinion shapers, or the government.

This single technical development has completely unhinged the status quo, resulting in a mad scramble to circumscribe and contain it, even as the Powers That Be, with each passing day, become more dependent upon the Internet themselves.

And just as the Internet shows up and blasts wide the all but hopelessly blocked information channels, so also is something similar happening to the human race. Despite an immense systematic campaign made to blight our health, limit our potential, and keep us down and docile, countervailing forces can be found suddenly popping up all over the place.

Take a look at the American diet, as expressed in fast-food menus. Where burgers, fries, and the like once reigned supreme, we now find chicken, salads, yogurt, and even fruit bowls. Nor is this all. Around the world, what may fairly be called superchildren are being born, children whose development cycle is so accelerated it is breathtaking and who are intellectually far beyond the existing norms. Likewise, despite unending efforts through means fair and foul to both limit the birth rate and increase the death rate, the birth rate on this planet continues to increase, building toward some sort of critical mass, perhaps. So, what happens to humanity when it's reached?

Arthur C. Clarke (now Sir Arthur) addressed this matter in his science-fiction classic *Childhood's End,* in which superchildren were born, grew up, and—having nothing in common with their progenitors—left the planet, leaving the entire global society, deprived of both a future and a reason for being, to implode. Earth today is going through profound changes, too, but not only from the "global warming" we're told about. Instead of atmospheric heating raising the temperature, what seems to be happening is that the internal temperature is rising, apparently, as the result not only of anomalous solar bombardment but also, possibly, because of energies linked to the stars themselves. Though the public may be carefully kept in the dark, the frequency and severity of earth's "natural" upheavals have gone up dramatically. Certain metaphysics texts indicate that Mother Earth is literally pregnant and about to give birth (calve a satellite?), and there are earth sensitives, both female and male, known to the writer whose condition reflects this, though they are not pregnant. Indeed, they report terrible labor symptoms.

THE LINK TO RETROTRANSPOSONS

What has all this to do with retrotransposons? Plenty, it turns out. Did you know that scientists have begun to link so-called starquakes (stellar internal upheavals) with subsequent earthquakes here? Supporting the theory that light can cause fundamental biological changes, as far back as August 8, 1998, *New Scientist* reported on page eleven: "Last year, astronomers showed that the Orion Molecular Cloud—adjoining the Orion Nebula—contained circularly polarized light that preferentially destroyed right-handed amino acids."

And this is without discussing such exotica as the effect of earth's passing through the photon belt/Manassic belt or emerging conclusions regarding what some call the Great Central Sun at our galaxy's core. What it all boils down to, though, is light and energy. And it is now clear that light can have profound impact.

According to Laura Knight-Jadczyk, she was told, "Study supernovae." She did and in subsequent questioning learned that the Crab Nebula (a supernova from five thousand years ago whose light reached earth nine hundred years ago) had the effects of "excitation of base liquid molecules" and "growth," growth and change in the size of people and in the psychological or mental sense. She further learned that there are superluminal effects that occur when a supernova is born, as well as effects linked to the physical arrival of the light from the event, and even things later than that. The real magic, "genetic splice of a strand," however, happens when a supernova is within 2,000 light-years of earth, and it so happens that both Betelgeuse and Rigel, in constellation Orion, are prime supernova candidates and in range, being some 1,500 light-years away. Elsewhere, we learn that supernovae allow dimensional "doors" to other universes to be redirected and that human beings once possessed 135 pairs of chromosomes (we now have 23) and that as far as regaining or restructuring our badly damaged DNA, what "was there will be again."

Fascinating!

So, light can and does cause fundamental biological changes. A mechanism for instantaneous, massive DNA activation, via retrotransposons, exists. An ET source directly links supernovae to both instantaneous and delayed wholesale genetic restructuring and repair.

ENDGAME?

Might the control game's goal, then, be so totally to trap us in this reality that we cannot respond to the coming stellar triggering (with the activation/repair of our entire genome) and ascend from this dimension to the next? Could that be why we are constantly beset? Anyway, it seems worth considering.

PART SEVEN

PARANORMAL POSSIBILITIES

29 **Paranormal Paratrooper**

For David Morehouse, the Military's Psychic
Warfare Program Was Much More Than an
Exotic Experiment. It Was His Mission, yet It
Nearly Destroyed Him.

J. Douglas Kenyon

According to the public information operatives of the CIA and the Pentagon, any official attempt to develop and deploy so-called psychic forces for military purposes was, at best, experimental, short-lived, and of no demonstrable value. Nevertheless, when the story broke in December of 1995 that—whatever the limitations—the Pentagon had resorted *at all* to "clairvoyant methods," it produced a spasm of disbelief in the world of paranormal debunkers. After all, hadn't it been shown to the satisfaction of any "rational" person that such stuff belongs entirely in the dark realm of witch doctors, mumbo jumbo, and superstition, or what the late Carl Sagan might have called "the Demon Haunted World"? It certainly does not belong—they would argue—in the arena of objective discourse. How could the hardheaded U.S. Defense establishment, with access to the best that academic science can produce, take such notions seriously—even for an instant?

From the initial stories, the CIA seemed more than forthcoming regarding its boneheadedness. According to the *New York Daily News* the spy agency had decided that, after "20 years and $20 million in tax dollars," the use of psychics and remote viewers (people capable of observing events at distant sites) "has not been shown to have value in intelligence operations" and "is not justified." All of which begs the questions: If the program was so useless, why did the CIA keep it around for twenty years? And why, after all those years of secrecy, did the agency finally decide to concede publicly its fruitless use of the taxpayers' money? The answers to these, and many other intriguing questions, may soon be coming to a theater near you. They are already available at your local bookstore.

Psychic Warrior, by David Morehouse (St. Martins Press, New York), tells with credible and vivid—almost excruciating—detail the story before, during, and after the author's years working in the Army's project Sun Streak (also known as Star Gate) at Fort Meade, Maryland, in the late 1980s and early 1990s.

A much decorated infantry officer and airborne ranger, Morehouse showed no signs of clairvoyant abilities in his early years in the Army, nor did he wish to. His sole desire was to pursue a traditional patriotic military career path. But all that changed when, during a training mission in the Middle East, a stray bullet struck his helmet, nearly killed him, and, in the process, triggered a dramatic alteration of some kind

Fig. 29.1. Former CIA remote viewer Dr. David Morehouse. (Photograph courtesy of David Morehouse)

in his mental machinery. Out of the nightmares and visions that followed grew a mysterious ability, which the Army was ultimately to harness for its sophisticated efforts—with the aid of paranormal techniques—to solve the very real problems of military intelligence. In his book, Morehouse reports not only on the extensive and effective use of remote viewing, in which he participated, but also on the internal political climate surrounding it, and the almost unbearable pressure to which he and his family were subjected before he was finally able to escape the Army's clutches and go public.

The dramatic potential of his story has not escaped Hollywood. A forthcoming major motion picture version of the book reportedly will star Sylvester Stallone, Morgan Freeman, and Kurt Russell. For the general public, however, the most important revelations include a highly credible description of an extremely unconventional program that, if it exists at all, is a direct contradiction of some of the most fundamental assumptions in the basic worldview of materialistic Western science.

For those who follow these issues closely, the story is not entirely news. Reports of such programs originally surfaced in the 1970s, when breakthroughs in the use of the paranormal were claimed by two physicists at the Stanford Research Institute in California. Russell Targ and Harold E. Puthoff

Fig. 29.2. Morehouse (second from left, top) and his troops "looking into" El Salvador a few months before being "shot in the head." (Photograph courtesy of David Morehouse)

in their book *Mind-Reach* reported on the ability of many people, using a technique that they called "remote viewing," to observe actual happenings at distant sites. Their principal subject, a New York artist named Ingo Swann, provided astonishing details about faraway locations with which he was previously unfamiliar. In one celebrated instance he gave an accurate description of Kerguelen, a remote island in the Indian Ocean. Subsequently, the program was made secret and dubbed Project Scannate by the government, and, as Morehouse told *Atlantis Rising* in an interview, most of the really great research, which truly proved the phenomenon, is still classified. But—official downplaying notwithstanding—it is clear that whatever originally attracted the military to the possibilities of remote viewing, it still had not entirely lost its appeal twenty years later. That is—or so they would have us believe—until December 1995.

According to Morehouse, however, the motive of the intelligence community in making its much publicized psychic warfare concession was nothing more than damage control: an attempt to put an official spin on the inevitable revelations that they knew to be forthcoming from Morehouse and others. In fact, one television special on the Discovery Channel had already provided a complete report. Far from being a poorly funded and unused technique, remote view-

Fig. 29.3. Physicist and zero-point energy advocate Harold Puthoff.

ing was an important and frequently used tool for serious intelligence analysts with a far larger budget than was stated. Though, as Morehouse is quick to point out, it certainly is not perfect.

The technique is 60 to 80 percent accurate, he says. "It's never 100 percent accurate. Never has been. Never will be." Nor is it ever a stand-alone endeavor. No mission was ever planned based solely on the report of one remote viewer. "Any operational target (someplace the program manager wants to know more about) is worked by a number of remote viewers. (Viewers are not allowed to discuss their work with each other.) All of the data is compiled and compared and presented in a final document to DIA (Defense Intelligence Agency)." The report is always used "in consonance with other intelligence collection methodologies—whether photo, signal, human, and so on. It is only one piece of the jigsaw puzzle." The ability of each remote viewer, Morehouse explained, is regularly tested and graded against known targets and a virtual batting average is thus determined. The management knows who is hot and who is not. When all the data are combined and statistically analyzed, the composite can be rated for accuracy in a way that makes it quite useful for inclusion in a comprehensive intelligence report.

"No intelligence product is 100 percent accurate," he declares. "A satellite photo of a top of a building in the Soviet Union does not tell you what's in it."

Unlike other forms of intelligence gathering, though, remote viewing puts no agent in harm's way, and requires no physical apparatus. (Morehouse concedes that to impress visitors at the Fort Meade office of Sun Streak, the staff had maintained—in the best tradition of science fiction—a dentist chair and a battery attached to a box.) Some viewers, including Morehouse, found it useful to employ biofeedback equipment, to help in achieving the proper brain wave state (theta), but others used no physical paraphernalia at all.

In the book, he recounts how he was trained (not by Ingo Swann, he told us, though many other remote viewers were) and how he learned to follow coordinates hidden in sealed envelopes and to navigate through what he calls the ether to a designated target. He details training and operational episodes in which he would travel, out of his body, to such places as the aftermath of a helicopter crash in Central America, or onboard Pan Am 103 before, during, and after the terrorist-caused explosion that destroyed it (that, he says, is how clues were generated which put investigators on the right track), and even into the oil fires following the Gulf War and to many other locations inaccessible by conventional means. He tells of investigating drug-smuggling ships during the war on drugs, counterterrorist operations (like the search for Lieutenant Colonel Higgins in Lebanon), and even visits to other planets, such as Mars.

Ultimately the experience was to convince him that remote viewing was far too valuable to the human race to be the exclusive property of the defense establishment, and he began to prepare a book. Shortly afterward he found himself unexpectedly threatened with court-martial on several seemingly unrelated and apparently trumped-up charges.

All of the charges made against Morehouse were based on the unsubstantiated allegations of a woman in what writer Leonard Belzer called a "92-page diatribe that was obviously designed to cover the gamut of potentially damaging finger pointing." The Army, it seemed, was determined to discredit him.

The ordeal that followed nearly cost Morehouse his sanity and came close to destroying his family, but ultimately he was able to leave the Army—though under less than desirable circumstances—to recover and to write his much delayed memoirs. The CIA's "Stargate" announcement, Morehouse says, followed by eight days the *Hollywood Reporter*'s item on the signing of his book and movie deal.

Remote viewing is not the only so-called paranormal phenomenon that U.S. and foreign intelligence agencies have attempted to exploit. Morehouse's own training included "map dowsing," a method of identifying secret locations by using assorted paraphernalia and maps. Other techniques he learned about indirectly. The most explosive possibilities have to do with something called "remote influence."

CIA and DIA officials deny emphatically that the government has ever attempted to use psychic means to influence anyone. Morehouse counters that Uri Geller—the famed Israeli psychic, a friend of his, who was an

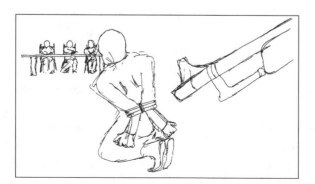

Fig. 29.4. Morehouse's drawing following a remote viewing session seeking information on Marine Lieutenant Colonel Higgins, who had been kidnapped by Lebanese terrorists. (Image courtesy of David Morehouse)

operative for the Israeli intelligence service, Mosad, before being sent to the United States to work with Targ and Puthoff on the original Scannate project—claims to have participated in such attempts. Geller has recounted on national television how he was used at a treaty-signing ceremony to influence targeted officials by psychic means. Geller has produced a photo of the event that includes him and the vice president at the time.

Many startling revelations of government dabbling in the black arts for purposes both foreign and domestic could be forthcoming if the principal operatives fail to maintain the secrecy that has sheltered their activities. Morehouse suspects that his own unit at Fort Meade was actually involved in trying to identify so-called double agents, who were in fact U.S. political figures. He cites a frenzy of document-shredding that preceded an official inspection of the organization. "There was an IG [Inspector General] directive" he recalls, "and a great deal of shredding went on and nobody was asked to desist, and we [Morehouse and his Sun Streak colleagues] don't know what it was that was being shredded, but it was hundreds and hundreds, if not thousands, of documents."

Psychic Warrior is not Morehouse's first effort to make his story public. An earlier collaboration, with Jim Marrs, author of *Crossfire: The Plot That Killed Kennedy* (on which Oliver Stone's movie *JFK* was based), was dropped by the publisher when one of the principal sources backed out at the last minute. Marrs's research included extensive taped interviews with, among others, Ingo Swann, in which Swann admitted that, in addition to training the many remote viewers with which he is credited, he trained another special group with a much more mysterious purpose.

Whatever the story of U.S. involvement in the paranormal, it is clear that the Soviet Union and others were there first. "The KGB pumped millions and millions of their money into their paranormal research program," Morehouse says, "It was there for decades before we ever picked it up and entertained it, and they were very practiced and very good. As were the Czechs, the Chinese, and the Israelis." Morehouse says he has seen pictures of remote-viewer warning detectors used in the Kremlin. He's also seen what he believes to be convincing evidence that the KGB developed and manufactured psychotronic weapons (capable, it has been alleged, of broadcasting such things as disease patterns and earthquakes), which, in the wake of the fall of the Soviet Union, are presently unaccounted for.

The strange fate that would bring Morehouse to the point of knowing and proclaiming such things has come as a complete surprise to him. Nothing in his earlier life foreshadowed it. "I've never been involved in self-hypnosis or trying to achieve out-of-the-body experiences, nor into ancient religions or any of that kind of stuff," he claims. "I'm just an infantryman who was shot in the head and began having very bizarre experiences." All which he believes helps his credibility. His attitude toward all that he has observed has been skeptical from the start. "The biggest criticism they had of me in the program the first year I was there was that I didn't even believe my own work," he recalls. "I mean I could always find some way of rationalizing why I had come to the conclusions that I had come to, or obtained the data that I had obtained." Being a skeptic—which he defines as "asking intelligent questions and expecting intelligent logical well-formed answers"—is good. "But being close-minded is not good. That's just a sign of ignorance. Unfortunately there are a lot of people who choose to be close-minded."

Some, apparently, choose to be more than close-minded. They choose not only to deny, but also to suppress the truth and to try to silence those who speak it. For those, Morehouse has taken special precautions. In his possession are documents detailing the real use made by the government of remote viewers in the "Operation Rice Bowl" debacle—the ill-fated Iranian hostage-rescue attempt in the last days of the Carter administration. For his own protection Morehouse has dispersed many copies of those documents throughout the country to be released if necessary.

In the meantime, he continues to try to sell anyone who will listen on the immense value that remote viewing can have for humankind. Credibility for the technique, he believes, depends upon recognition not only of its possibilities but of its limitations as well. Ultimately, though, he believes, its application to society's problems such as crime solving, resource location, and even industrial espionage is inevitable. Recently he's been asked to lend his talents to the search for clues to the fate of TWA flight 800.

Another area where a good remote viewer could make himself useful would certainly be archaeology. Morehouse is already on record in his book as having seen the Ark of the Covenant in its present hiding place. Where, in fact, is it now, anyway? "Well, it's no place we're going to find it," he says with a laugh. "It's not in North Africa, and I'll tell you what, I don't believe those monks are responsible for it [a reference to Graham Hancock's Ethiopian theory]."

Once again, the plot thickens.

30 Psychic Discoveries since the Cold War

Sheila Ostrander and Lynn Schroeder Revisit the Revolutionary Research of Their Classic Study

Len Kasten

In June of 1968 Sheila Ostrander, a Canadian, and Lynn Schroeder, an American, were invited to attend an international conference on ESP in Moscow. The invitation was from one of the most fervent missionaries of Soviet psi research, Edward Naumov, then thirty-six. A few years earlier, when overt interest in such subjects could easily result in a long holiday in Siberia, such a conference would have been impossible. Then suddenly, in the mid-sixties, the doors of prohibition in Russia, under the "Troika" rulership, clanged open. Ostrander and Schroeder, encouraged but still doubting the turn of events, started writing to Soviet scientists and researchers. For three years, the letters and packages describing Soviet psi research poured in, culminating in Naumov's invitation to "come and see for yourself."

They did, also visiting Bulgaria, but even before the conference had ended the suppression returned. The conference was closed down and Ostrander and Schroeder had to take refuge in Prague. Once again they were one step ahead of a Soviet crackdown, getting out of Prague only days before the Soviet tanks rolled in. During that brief lapse in Soviet repression came an epic book

Fig. 30.1. Authors and researchers Sheila Ostrander and Lynn Schroeder. (Photograph courtesy of Sheila Ostrander and Lynn Schroeder)

that opened the eyes of the world to the astounding breakthroughs in psychic research in the communist countries. That book, *Psychic Discoveries Behind the Iron Curtain,* published in 1971, became an instant underground hit in New Age circles, and even though it never achieved bestseller status in the mainstream readership, it nevertheless has now become something of a classic.

Up to that point, all we had to go on was some very tentative research by Dr. Rhine at Duke University in the fifties. While some of Rhine's conclusions were positive and dramatic, the effect was blunted because the research results were couched in cautious, dry statistical terms, and consequently it was difficult to appreciate the real impact. But the publication of *Psychic Discoveries* made it all very real. The future of the human race became fantastic to contemplate, and at the same time very chilling. The authors suggested that though these discoveries could lead to a Utopia if properly utilized, they could also lead to a hell on earth if abused.

The authors have become world authorities on these subjects and are in great demand. *Atlantis Rising* was fortunate to have the opportunity to interview them at John White's 1997 UFO conference in New Haven, Connecticut. With publication of their new book by Marlowe & Company in New York (paperback), the pair are back in the public eye now. Basically an update of the original, the new volume bears the shortened title *Psychic Discoveries*. Publication of the new edition was motivated primarily by the flow of formerly top-secret information coming out of the Soviet Union since the end of the Cold War. The new book contains an abridged version of the old classic, and then a second part entitled "Psychic Discoveries—The Iron Curtain Lifted."

Far from idle since 1971, the authors published *Supermemory* (New York, Carroll and Graf) in 1991. This book, which evolved out of contacts and interests developed while researching the original *Psychic Discoveries*, bids to become a classic in its own right.

The revelations in all three books are nothing short of sensational, yet for over twenty-five years, the press and the public have barely noticed. Echoing the pattern found with the UFO phenomenon, some believe the situation may suggest a world cover-up. In fact, the authors told us that they have now recognized that UFO secrets and psi secrets seem inextricably linked. Uri Geller, after all, claimed to have obtained his powers from extraterrestrial sources. The new book includes a foreword by Geller in which he marvels that the press has taken little notice. He mentions a press conference in 1977 at which Stansfield Turner revealed that the CIA had a parapsychology program in place, and had found a man who could see through walls (Pat Price). The revelation created not even a ripple in the media!

Yet while the revelations in *Psychic Discoveries* did not get wide publicity, they were nevertheless revolutionary. The impact on society has yet to be fully appreciated. The discoveries of an obscure electrical repairman from the Black Sea city of Krasnodar were first revealed to the West in *Psychic Discoveries*. In classic Ostrander-Schroeder style, the drama surrounding the experimentation of Semyon Kirlian and his wife, Valentina, is brought to life. It was in this chapter where the terms "energy body" and "bioplasmic body" were first used and the idea of the aura was first put forth. One of the most significant results was a new understanding of the ancient Chinese practice of acupuncture. A Russian surgeon, Dr. Mikhail Gaikin, showed that the colored lights erupting from the body and appearing in the Kirlian photographs were actually coming from the seven hundred acupuncture points.

Before the publication of *Psychic Discoveries*, there had been several books written about the unique and strange dimensions of the Great Pyramid of Giza speculating about their significance. But thanks to their side trip to Prague, Ostrander and Schroeder were the first to tell the world about "pyramid power."

It was there that they were introduced to Karel Drbal, a Czech radio and television engineer, who had discovered that small pyramids of the same relative dimensions as the Great Pyramid could sharpen razor blades! The pyramidal shape apparently focuses cosmic energy when precisely aligned on the north–south axis, which can renew the crystal line structure of good-quality steel.

MIND WARS AND SOCIAL CONTROL

Without exception, all of the Soviet researchers interviewed hoped that these discoveries would be used only for good, but clearly recognized that many of them offered potential in intelligence and counterintelligence, and that some could be used to make very destructive weapons—and so did the CIA.

Although we now know that U.S. intelligence agencies have been conducting clandestine, black-funded psychic experimentation for many years, there is considerable evidence that many of the top-secret government psi programs were triggered by *Psychic Discoveries*. Extensive U.S. programs to monitor the Soviet research started up about that time. The authors were invited to speak to the new Congressional Clearinghouse of the Future, and parts of the book were read into the *Congressional Record* by Congressman Al Gore, who later became the chairman of the committee, and who has maintained a high degree of interest in psychic matters.

Apparently during the Cold War, both sides were in a no-holds-barred race to perfect psi weapons, but, just as with space programs, they may now be cooperating. From the original version of the book, the world first learned of an astonishing Soviet development: the ability to control behavior and consciousness telepathically! In the chapter entitled "The Telepathic Knockout," the authors reported on experimentation dating from 1924 in which Soviet scientists successfully placed subjects in hypnotic trances and awakened them telepathically across thousands of miles. Once the connection was established, a subject's behavior could be manipulated by suggestion, just as in face-to-face. Typically, they can carry on conscious conversation and activity while in the trance. In the new book we learn that the CIA has picked up this ball and run with it.

But it was from the Czechoslovakians at the conference that the authors learned of a discovery that promises to ultimately make twentieth-century explosive weaponry seem as primitive as the horse and buggy: the psychotronic generator. And while in Prague they met the inventor, Robert Pavlita, a design director for a large Czech textile plant. In a documentary film produced by a major Czech studio, the authors saw small, strange-looking metal objects that appeared to be designed by Picasso arrayed on a table. They had 110 moving parts. In the film, Pavlita explained that the secret was in the form. "The generators accumulate human energy," he said. Then they focus this energy to carry out various types of work. Pavlita and his daughter Janna charged the generators by gazing at them in a staring pattern. Once charged, they turned rotors, attracted nonmetallic particles, caused seedlings to grow larger plants, and purified polluted water. This human, psychic energy has had many names since ancient times, variously referred to as prana, chi, vital energy, animal magnetism, odic force, etheric force, orgone (Reich), and now bioplasmic and psychotronic energy. At Pavlita's home in Prague, the authors handled the devices and saw personal demonstrations by the inventor himself. But what happens when a psychotronic generator is pointed at a human? Pavlita's daughter volunteered to be a guinea pig. She became dizzy and lost her spatial orientation. The device can also kill flies instantly.

In the new edition we learn that former KGB Major General Kalugin started talking in 1990. He claimed that Yuri Andropov gave orders to move full speed ahead with psychic warfare in the early seventies, and obtained funding of 500 million rubles. The Soviets then developed sophisticated Pavlita-type generators. Dr. Nikolai Khokhlov, a Russian CIA operative, uncovered over twenty heavily guarded, well-funded laboratories working on psychotronic devices for military use in the seventies. Some of this effort may have been cooperative with the United States.

THANKS FOR THE MEMORIES

At least one U.S. psi program that we know of predated the publication of *Psychic Discoveries*. In *Supermemory*, the authors reveal details of the CIA EDOM program. EDOM means Electronic Dissolution of Memory, and apparently this technique was perfected years ago. The CIA can zap long-term memory and turn someone into an amnesiac zombie by blocking the neurotransmitter acetylcholine and by electronically interfering with the bioplasmic body. Apparently, they routinely employ this technique to "neutralize" former top-secret operatives, just as in the movie *Total Recall*. This capability was developed under the MK-ULTRA program when they performed bizarre memory experiments on mental hospital inmates, prisoners, and research volunteers in the 1960s, before the program was halted by Congress in 1976.

Perhaps the most bizarre development in memory control was inspired by multiple personality disorder. Also in *Supermemory*, we learn that the CIA can artificially seed multiple personalities in the same body, each with its own memory bank not accessible to the others. Gil Jensen, an Oakland, California, CIA doctor, claimed that he created a personality named Arlene Grant in the body of famous supermodel Candy Jones in the fifties and sixties using hypnosis and memory-altering drugs. Grant was trained as a super-spy and given a complete memory history and top-secret information, which Jones knew nothing about. Whenever Jones went on celebrity trips, Grant was summoned on the telephone through a series of electronic sounds and carried out her spy missions. The primary personality can never reveal information from the secondary memory bank, even under torture, and therefore makes the perfect spy. This program is now called Radio-Hypnotic Intra-Cerebral Control and is apparently based on Soviet discoveries related to electromagnetic manipulation of the bioplasmic body.

UFO SECRETS REVEALED

By far the most sensational revelations coming out of the post–Cold War Soviet Union are concerned with UFOs and the moon. In the new edition we learn that information from now opened KGB files tells of widespread UFO sightings reported to the Soviet military in the years following the incredible Voronezh public landings in September 1989, which made headlines all over the world. Hundreds of adults and children saw the spacecraft landings and giant aliens and small robots moving freely about downtown Zavodsk Square in broad daylight. Thousands more saw giant disks hovering over the city's nuclear power plant.

According to the KGB files, in March of the following year, over a hundred military UFO observations were reported to the Air Defense Forces, and a 300-foot disk hovering over the headquarters of the Soviet Air Defense Command was reported in April. Also in the KGB files, according to the Hungarian Minister of Defense, George Keleti, a former colonel in the army, UFOs swarmed over Hungary at the same time as the Voronezh landings, and alien craft landed at military air bases all across the country. Keleti claimed that the four-foot "robots" actually attempted to climb into Hungarian MIGs and repelled guards with ray guns! Then the ten-foot humanoids became invisible when fired upon with machine guns. According to *Psychic Discoveries,* an avalanche of formerly "concealed sightings, landings, close encounters, abductions, and more" has seeped out to the press and to newly formed UFO groups in Russia since 1990.

But the secrets eking out of the Soviet space programs are even more exciting. Soviet Air Force Colonel Marina Popovich showed photos of a fifteen-mile-long object flying near the Martian moon Phobos, taken by the Soviet probe Phobos-2, at a conference in San Francisco in 1989. Russia's Luna 9 moon probe, which landed in the Ocean of Storms on February 4, 1966, took some spectacular three-dimensional photos over the Sea of Tranquility, which showed a group of spires that appeared to be obelisk-shaped, and were obviously artificial structures. Soviet space engineer Dr. Alexander Abramov subjected the photos to a complex mathematical analysis and concluded that they were archaeological ruins. Furthermore, he told the authors that the obelisks on the moon were arranged in exactly the same pattern as the pyramids of Giza when placed on an *abaka,* an ancient Egyptian grid of forty-nine squares. Then Dr. S. Ivanov, one of Russia's most eminent scientists, published an analysis in the Soviet magazine *Technology for Youth,* claiming that the monuments were arranged according to definite geometric laws, and were evidently "artificial structures of alien origin."

U.S. photos taken by Orbiter 2 on November 20, 1966, tended to confirm the Soviet findings. Dr. Ivan T. Sanderson, science editor of *Argosy* magazine, analyzed these photos and claimed that the tallest structure was

Fig. 30.2. A detailed analysis of an obelisk complex on the moon photographed by a Russian space probe. This photograph appeared on the cover of the Russian publication Technology for Youth. *The largest spire stands fifteen stories high. (Image courtesy of Sheila Ostrander and Lynn Schroeder)*

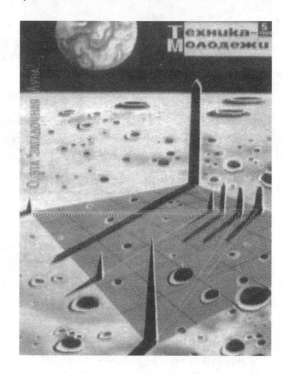

about fifteen stories high and the smallest about the size of a fir tree. The authors later found out that NASA had classified hundreds of lunar photos and still refuses to release them.

We conclude with a succinct and eloquent summation of the situation by former astronaut Dr. Brian O'Leary, as quoted in *Psychic Discoveries:* "The cosmic Watergate of UFO, alien, mind control, genetic engineering, free energy, antigravity propulsion, and other secrets will make Watergate and Irangate appear to be kindergarten exercises . . . but, the truth will and must be known eventually."

31 | When Science Meets the Psychics

New Research Buttresses the
Challenge of Mind to Matter

Patrick Marsolek

Behind closed doors in science, business, and academia, clandestine meetings are taking place.

A large university has problems with its computers. Normal troubleshooting procedures don't solve the problem. A person considered to possess extrasensory perception, usually referred to as a "psychic," is called in. She intuitively locates a break in a sealed cable where one was neither seen nor suspected. In another case a doctor sends samples of blood, hair, and saliva to an intuitive for help in diagnosing a patient's illness.

Professionals generally take great care not to discuss these developments in public. They do not want to be ridiculed. So why do they do it? Because in many cases, it works. Out of the sight of the public, intuitives are being hired by topnotch companies to do all sorts of things, from locating geological faults in the earth to helping to predict earthquakes, from helping physicians to diagnose illnesses to financial forecasting to assisting corporate managers to make business decisions.

The cutting edge of science is moving ever closer toward matters of the mind and subjective experience, as seen in such areas as field mechanics and quantum physics. Researchers are probing possibilities and realities that in an earlier era would have fit better in the *National Enquirer* than a scientific journal. Science is edging closer to Einstein's unified field theory and a world that is increasingly being seen as made of consciousness rather than matter.

A revolution is afoot in the scientific world concerning subtle energies that seem to take two different paths—and meet in the middle. The first is represented by a large number of people who are beginning to recognize the value of intuition in their everyday lives and who actually use intuitives. This is what could be called the working path. On the other path, the path of knowledge, are scientists in active research and investigation on the leading edge of science, including areas like chaos theory, field theory, and the aforementioned quantum physics. Many of these leading-edge fields are finding themselves forced to include consciousness and its interaction with the world.

An example of the working path is Raymond Worring, director of the Investigative Research Field Station in Helena, Montana. Worring, coauthor of *Psychic Criminology,* has spent thirty years developing an extensive body of knowledge on using intuitives for practical purposes. He's worked with several hundred psychics from all walks of life on murder investigations, finding missing persons, and archaeological investigations.

Worring and co-author/researcher Whitney Hibbard have worked extensively with George McMullen, one of the foremost intuitives operating today. McMullen, author of *One White Crow, Red Snake, Running Bear and Two Faces,* has worked on Native American archaeological sites in North America and Canada. McMullen joined forces with the late Canadian archaeologist and anthropologist Dr. Norman Emerson from the University of Toronto. He assisted Emerson in revealing the location of ancient Iroquois and Huron villages, including the remains of longhouses, the post-hole indications of palisade fortifications, and ancient burial sites. Dr. Emerson risked his reputation by endorsing intuitives in archaeology and anthropology. He wrote in a paper to the Canadian Archaeological Association, "By means of the intuitive and parapsychological a whole new vista of man and his past stands ready to be grasped. As an anthropologist and as an archaeologist trained in these fields, it makes sense to me to seize the opportunity to pursue and study the data provided."

Fig. 31.1. Author and intuitive George McMullen, who employs his gifts in the fields of archaeology and criminology.

Since he started with Dr. Emerson, McMullen has put his archaeological skills to use at sites around the world, including Ecuador, Israel, Egypt, and many sites in the United States. One trip with Hugh Lynn Cayce was made to try to confirm Edgar Cayce's readings indicating the existence of a "hall of records" on the

Giza Plateau. Some of the information that McMullen conveyed was that "the Sphinx had a crown on it at one time and that they would find the hall of records where the crown of the sphinx made a shadow at sunrise on the ground."

Worring and Hibbard also have worked with McMullen in the area of criminology in conjunction with several law enforcement agencies. McMullen's skills were applied to murder investigations and missing persons. Worring and Hibbard's research with McMullen and many other intuitives led to the publication of *Psychic Criminology,* a training manual for the use of intuitives in law enforcement. A private detective in St. Louis, Missouri, Rich Brennen, explains the pragmatic value of intuition: "I use psychics, the pendulum, and remote viewers as tools of my profession. It is just like any other investigative tool, like computer databases and video surveillance . . ."

Worring and Hibbard also collaborated with intuitive Francis Farrelly on criminal and archaeological cases and specifically for developing intuition as an investigative technique for private detectives. Farrelly's long career has included using intuition in criminology, medicine, computer troubleshooting, archaeology, and stock market predictions. She worked as an agricultural consultant in using radionics, the controversial technique developed by Dr. Albert Abrams, which utilizes the radiations specific to individual organisms to effectively curb spruce budworm infestations.

Another psychic who is drawing considerable attention is Annette Martin, the "radio psychic" from the San Francisco Bay area whose intuitive work includes diagnosing disease and conducting psychic hearings, assisting in criminal investigations, "ghost busting," and consulting with corporations, such as Hughes Aircraft and Sun Microsystems. She also trains psychics at her Institute of Intuitive Research in Campbell, California. Both the Discovery and History television cable channels aired documentaries about her in 1998.

While some intuitives use their abilities in a variety of ways, some excel in specific areas. Some psychics will specialize in one type of victim, such as a dead body, a drowning victim, a missing person, or a missing animal. Though other intuitives, such as Uri Geller, Ingo Swann, and Ron Warmouth, seem to have wide ranges of skills, they excel when they use their talents for prospecting oil and precious metals.

Geller runs a business that also consults with engineers and geologists to do mineral prospecting. Beverly Jaegers and her group, US Psi Squad, based in St. Louis, Missouri, work extensively in criminology and missing persons. Jaegers is also conducting trainings in remote viewing and develop-

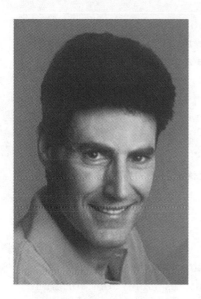

Fig. 31.2. World-renowned intuitive Uri Geller.

ing psychic abilities. Her group currently is building a formal network of detectives/police/psychics to work together in crime solving.

Despite the fact that these nonverifiable phenomena are largely unacceptable to mainstream science, more and more professionals from many fields are seeking the help of intuitives. Attorneys use intuitives in negotiations; high-level executives turn to them for help in management and decision making, financial forecasting, and detecting problematic situations before they arise. Not only are professionals working with intuitives, but many are also pursuing training for themselves in order to enhance their own intuitional capabilities.

The use of intuitives is still highly controversial. Many professionals will use psychics only as a last resort, while keeping it a secret if possible. Many psychics also desire confidentiality as a result of harassment they've experienced. One conservative group has printed pictures of psychics on posters alongside neo-Nazi propaganda, labeling them "devil worshipers." Annette Martin says that although negative responses in her case are low, she nevertheless tries to "keep a fairly low profile with police work"; this helps to gain agencies' confidence in her work and provides a buffer of protection from extreme opponents. As an added complication, organizations have had to learn to be careful about utilizing psychics whose main or only interest is publicity.

When intuitives are called in, they are often not acknowledged for their help, especially if they are working in tandem with other, more acceptable procedures. While collecting information for this article, most psychics and investigators whom I contacted would not divulge specific names and circumstances due to "restrictions of confidentiality." The police chiefs, administrators, and doctors whom they had worked with would find their jobs at risk if it were generally known they had utilized these valuable resources, in spite of the sometimes remarkable results that were obtained. In one case the practitioner was arrested when his results were too successful and threatened the structure of the system.

The U.S. government's documented use of intuitives in the area of remote viewing has helped considerably to bring intuition into the mainstream. There are many opportunities to learn the methods of remote viewing; in addition to Bev Jaegers's group, there are also Ingo Swann, Lyn Buchannon, Major Ed Dames, Uri Geller, Angela Thompson, and the Monroe Institute, to name a few different groups each teaching its own version of RV. The popularity of remote viewing, with its structured protocols, may be one indication of greater acceptance of "psi-phenomena."

Virtually all of the world's cultures—with the notable exception of Western industrial civilization—have held these extrasensory channels of information in high regard and have spent much time accessing states of consciousness that facilitate the intuitive process. New scientific discoveries about the human psyche and body are revealing that a high degree of correlation exists between states of consciousness and the experience of reality, something many nontechnological cultures simply took for granted. This leads us to the path of knowledge that is drawing close to the realms of subjective experience.

Research into hypnogogia, the state of consciousness between waking and sleep that each of us experiences every day, is shedding light on the intuitional process. Andreas Mavromatis, in his book *Hypnogogia,* writes that intuitional experiences are distinguishable from hypnogogia only "by the subject's set of beliefs and the setting in which the experiences take place." Many scientists might be surprised to learn how often the intuitional process contributes to scientific insights. This similarity is illustrated by Thomas Edison's use of catnaps. When Edison reached a creative impasse he would take a nap, if only for a few moments, and often would have a creative insight relevant to his task at hand.

On the physical level, researchers recently have discovered cells in the brain containing magnetite, indicating an ability to sense energy fields. The proximity of these cells to the pituitary and pineal glands has led Richard Lawlor, author of *Voices of the First Day,* to propose that these glands may use information from earth's magnetic field to regulate the release of hormones in the brain that directly control levels of conscious awareness. Other research is indicating the evidence of organic crystalline structures in the body, such as the rhodopsin molecules in the cone and rod cells of the retina that are assembled in crystal-like plates.

The presence of these specialized organic structures within our bodies indicates an ability to interact with different kinds of energy fields that exist all around us in the atmosphere. Researchers exploring the frontiers of field theory such as neurochemist Glen Rein, at Stanford University, and the nuclear engineer Colonel Thomas Bearden, are revealing to the world the potentials of scalar fields, phenomena that may shed some light on what information field intuitives may be accessing. *Scalar* refers to a quantity that has magnitude or size but no motion. An example of a scalar quantity is pressure. The pressure of a gas has a value and we can measure it, but pressure doesn't include motion of that energy; it has no motion. When applied to field theory, *scalar* refers to fields of potential, energy, and information

Fig. 31.3. Thomas Edison in his laboratory.

that lie outside of the usual spectrum of electromagnetic energy. But this can be confusing, because scalar fields may be coupled with electromagnetic phenomena. Glen Rein has said, "Scalar fields . . . are distance and time independent (unlike electromagnetic fields); they act at a distance; they can have negative energy; they even have the characteristic of being able to travel backward in time!"

The highly controversial nature of these often undetectable fields is due largely to the difficulty of studying them. They exist, invisible and unseen, until triggered by some other form of energy, much like the holographic image appearing when a laser beam is applied to the photographic plate. Recognizing these scalar potentials around us in our environment may give validity to such theories as the zero-point energy postulated to exist within all matter in the universe, or to Rupert Sheldrake's morphogenetic fields created by living species, or even Jung's collective unconscious. The fields themselves are still mysterious and difficult to understand, but their existence is apparent from their undeniable effects, such as those of the gravitational and magnetic fields, and some might even say the phenomena of psychic information. Regarding the human experience of these fields, Glen Rein proposes a theory of "crystalline transduction" to explain the mechanism of action of scalar waves within biological systems. He suggests "that scalar energy is transduced into linear electromagnetic energy in the body by liquid crystals in the cell membrane and solid crystals found in the blood and biological tissues." These fields may be transmitting information by our very interaction with them.

Improved understanding of subtle energy fields and subjective states of consciousness is shedding some light on methods by which information is obtained outside rational or direct sensory processes. Science is beginning to understand the subjective experience of the intuitive. Science is ideas and ideas often originate in the same place for both the scientist and the psychic, in the medium of the higher mind, the quantum field, or higher consciousness, whatever term you want to employ. Once an idea is obtained, the rest is simply footwork—clarifying, substantiating, and exploring the possibilities. This is an essential process common to everyone, not just scientists or intuitives.

Humankind is poised at the start of a new millennium. One of the revolutionary benefits of developing a working understanding of these subtle energies in our lives lies in the transformation of our consensual belief systems regarding the structure of the world and our place in it. We may regard our world as the experimental physicist, Nick Herbert, proposes: "That all the stuff that physicists can explain is just a tiny amount of the real world. . . . There's a lot of mind, at least as much as there is matter, and we just aren't aware of it." Humankind is now grappling with ideas and information traditionally given religious sanction, using them to forge a new vision of what it means to be alive and creative. When I talked with Annette Martin, she said, "You would be amazed at how many people really believe that there is something beyond themselves and this beautiful planet we live on."

32 Rupert Sheldrake's Seven Senses

A Candid Conversation with a Scientific Iconoclast

Cynthia Logan

Ever looked up at a gaggle of geese and marveled at their exquisite, coordinated unity? Ever felt the urge to turn around and find that when you do, someone's eyes have been fixed on you? Both are common experiences that illustrate what cutting-edge biologist Rupert Sheldrake considers the "seventh sense." Unlike the sixth sense—which he says has already been laid claim to by biologists working on the electrical and magnetic senses of animals and is rooted in time and space—the term *seventh sense* "expresses the idea that telepathy, the sense of being stared at, and premonition seem to be in a different category both from the five normal senses and from so-called sixth senses based on known physical principles." Though the geese we gaze at have a built-in biological compass that enables them to respond to the earth's magnetic field, Sheldrake thinks there's more than magnetism afoot to keep them aloft and aligned.

The Cambridge-educated "heretical" scientist has, besides a love of plants and animals, a way with words, and has authored a number of award-winning books with subjects as intriguing as their titles. *Dogs That Know When Their Owners Are Coming Home, and Other Unexplained Powers of Animals* (1999) won the British Scientific and Medical Network Book of the Year Award, and *Seven Experiments That Could Change the World* (1994) was voted Book of the Year by the British Institute for Social Inventions. Sheldrake is also the author of *The Presence of the Past* (1988), *The Rebirth of Nature* (1990), and, with Ralph Abraham and Terence McKenna, *Trialogues at the Edge of the West* (1992) and *The Evolutionary Mind* (1998). His latest book, *The Sense of Being Stared At* (Crown, 2003), delves into just such senses—phenomena he asserts are worthy of investigation. "I argue that unexplained human abilities such as telepathy, the sense of being stared at, and premonition are not paranormal, but a normal part of our biological nature," he writes.

225

Sheldrake feels that prejudices rooted in the thinking of seventeenth- and eighteenth-century philosophers have inhibited research and inquiry: "If we only open our minds and make an effort to understand, we will be vastly rewarded with our new knowledge."

Not only does Sheldrake think we should be looking at heretofore unexamined phenomena but—and he feels strongly about this—we should bring science back to the people, to amateurs like you and me. Explaining that science is founded on the empirical method (on experience and observation), he turns to words again: "Billions of personal experiences of seemingly unexplained phenomena are conventionally dismissed within institutional science as 'anecdotal.' What does this actually mean? The word *anecdote* comes from the Greek roots *an* and *ekdotos,* meaning 'not published.' Thus an anecdote is an unpublished story." He points out that courts of law take anecdotal evidence seriously, often convicting or acquitting defendants because of it. He also cites its role in medical research, stating that "when patients' stories are published they are then elevated to the status of 'case histories.' To brush aside what people have actually experienced is not to be scientific, but to be unscientific."

For the past fifteen years, Sheldrake has focused his scientific interest on how systems are organized, pioneering what he calls the "Hypothesis of Formative Causation," consisting of "morphic fields" and "morphic resonance." Thanks to the Power Rangers and other kids toys, most of us casually use the word *morph* to mean "change into" or evolve. Precisely, says Sheldrake, whose work takes off from where the now widely adopted biological concept of "morphogenetic fields" (used to explain, for example, how arms and legs can have different shapes even though they contain the same genes and proteins) left off.

Sheldrake surmises that the fields evolve along with the systems they organize and coordinate. Since a field is a "sphere of influence," morphic fields would be those that can change or evolve their spheres of influence. He says there are morphic fields within and around individual cells, tissues, organs, organisms, societies, ecosystems, and so on—fields that are shaped by past events and patterns through an in-built memory called "morphic resonance." This is how, he reasons, instincts and "species specific" abilities develop; the enviable coherence of birds in flight is due to the morphic field that links them and the resonant memory that has evolved through millennia.

"'Instinct' is a vague term for an inherited pattern of behavior—conventional biology holds that instincts are programmed into the genes—I argue that genes are grossly overrated and don't do half the things they're cracked up to do," he quips. "What they do is code for the sequence of amino acids and proteins—they make the right chemicals." In *Seven Experiments That Could Change the World,* Sheldrake examines how termites structure their colonies and build arches, noting: "Of course the insects have genetic coding that prompts them to behave in certain ways, but the actual forming of the nest and the coordinating of the colony is accomplished with morphic fields." For those who find the concept hard to fathom, he offers an analogy: "Just as a magnetic field can influence the pattern of iron filings within it, so a morphic field can influence the behavior and movements of cells within a body or members within a group." He also maintains that such fields underlie the bonds that form between pets and their owners. And here is his love: pets and people learning from and helping each other.

With research assistants posted in a number of places around the world (London, Zurich, California, New York, Moscow, and Athens among them), Sheldrake maintains a large database of pet owners who have participated in "do-it-yourself" experiments that are simple but rigorous enough to produce evidence he can support. "Hundreds of videotaped experiments have shown that dogs are indeed able to anticipate their owners' return in a way that seems telepathic," he claims. Other results indicate that cats, parrots, homing pigeons, and horses are also highly telepathic.

Though some dismiss his views and conclusions as nonsense, Sheldrake certainly has the academic credentials to command serious consideration of his theories. He studied natural sciences at Cambridge and philosophy at Harvard, where he was a Frank Knox Fellow. Having earned a Ph.D. in biochemistry at Cambridge in 1967, he was a Fellow of Clare College, Cambridge, and was director of studies in biochemistry and cell biology there until 1973. As a Research Fellow of the Royal Society, he studied the development of plants (with an emphasis on the hormones within them) and the aging of cells, while enjoying seven years of stimulating academic conversation and luxurious accommodations.

"I lived in seventeenth-century rooms in a beautiful courtyard. At the ring of a bell, I would walk across the courtyard, put on my academic gown, and sit down at a table furnished with a delicious meal and vintage wine. After dinner we drank port in a paneled 'common' room and talked for hours," he recalls, adding that "since the fellows of the colleges are from different subjects, I had many valuable opportunities for interdisciplinary discussion."

The blend of academics and conviviality has served Sheldrake well:

As the author of more than fifty papers published in scientific journals, he accepts criticism without defensiveness, stating that "healthy skepticism plays an important part in science, and stimulates research and creative thinking." He differentiates an open-minded, healthy skeptic who is interested in evidence from a "skeptic," whom he defines as someone committed to the belief that paranormal phenomena are impossible. His extensive website (www.sheldrake.org) addresses specific comments from several skeptics. "Click on their names if you want to know what they said about my research on the unexplained powers of animals and to read my replies," he suggests. Though he's been ridiculed by some and teased by others ("some of my peers have suggested using telepathy instead of a telephone when I mention I'm going to make a call"), many other scientists find his conclusions fascinating and plausible. Quantum physicist David Bohm sees similarities in Sheldrake's "Formative Causation" and his own theory of an invisible "Implicate Order" behind the "explicate" material world.

At six feet two inches, Sheldrake appears lanky, thoughtful, and energetic. With vibrant eyes not shielded by glasses, he looks anything but the sci-nerd. Describing himself as being "in the middle" on the intro-extrovert scale, he looks like someone you'd approach at a party, someone with whom you would enjoy an animated, illuminating conversation. If you're lucky, he might hit the piano, playing, most probably, Bach. If you by chance play yourself, he would be delighted at the prospect of a duet. At home, he enjoys playing games with his sons, who have inherited his love of animals and participate in his experiments.

Fig. 32.1. English biologist and author Rupert Sheldrake.

As a youth himself, Sheldrake kept homing pigeons and was "always interested in plants and animals—they turned me on to biology," he says. "I was also quite interested in chemistry, partly from my own spontaneous interest and also from my father, a pharmacist and a keen naturalist. He had his own scientific laboratory at home and would do amateur microscopic research." Sheldrake's younger brother, his only sibling, is an ophthalmologist, a visionary in his own right.

Sheldrake's vision is "to be able to help open up the world of science so that phenomena presently ignored or neglected can be brought within the scope of science." He hopes that this "expanded" science will give us a better idea of the interconnections among ourselves, plants, and animals, and the planet as a whole, and between ourselves and the universe. "Such science would not be in conflict with spirituality, but complementary to it and could lead to a healing of the split between science and religion that has so damaged our civilization," he says. In his own life, that split has been healed, though it took some time.

Raised by devout Methodists in Newark-on-Trent, Nottinghamshire, England, Sheldrake attended an Anglican boarding school and found himself torn between a very Protestant tradition and the Anglo-Catholic "with incense and all the trappings of Catholicism." The rift between his love of living things and the mechanistic approach to biology he was taught was more intense. His discovery of an essay by the German philosopher Goethe about a holistic science that would not focus on reducing things to their minutiae and that would include direct experience and one's senses intrigued and inspired him. So much so that he spent a year's fellowship at Harvard ("where I found they treated me like a child!") studying the philosophy and history of science. "I read Thomas Kuhn's book *The Structure of Scientific Revolutions* and that had a big influence on me, gave me a new perspective," remembers Sheldrake.

From Harvard, it was back to Cambridge, where he did his graduate work and came across the Epiphany Philosophers, an eclectic group of philosophers, physicists, hippies, healers, mystics, and monks. "We lived together in a windmill on the Norfolk coast for a week at a time, four times a year, exploring new ideas in quantum theory, the philosophy of science, parapsychology, alternative medicine, and other sixties themes," he recalls. "We were a kind of vanguard."

From 1974 to 1978 Sheldrake was principal plant physiologist at the International Crops Research Institute for the Semi-Arid Tropics (ICRISAT) in Hyderabad, India, where he worked on the physiology of tropical legume crops, and remained consultant physiologist until 1985. For a year and a half, Sheldrake lived at the ashram of Father Bede Griffiths, a Christian Benedictine monk in south India, where he wrote *A New Science of Life,* considered his "magnum opus." But again, life wasn't all work and no play:

it was in India that he met his wife, Jill Purce. Both were speakers at the International Transpersonal Association's 1982 conference on "Ancient Wisdom and Modern Science": her lecture addressed ancient wisdom; his, modern science. The blend has been working for them ever since. The couple now live in London with their two teenage sons, three cats, a goldfish, and a guinea pig.

Like Dr. Candace Pert, the endocrinologist credited with the discovery of the opiate receptor in the brain, Sheldrake has concluded that the mind is not confined to the brain. While Dr. Pert focuses on the chemical proof that neuropeptides are found throughout the body, Dr. Sheldrake suggests that minds involve extended fields of influence that stretch out beyond brains and bodies entirely, connecting thoughts and intentions as well as creating the "memory" in nature. "I think morphic resonance works directly across time rather than being stored in a place, as on a CD or hard disk," notes Sheldrake, who finds the spiritual concept of akashic records to be "like an etheric filing cabinet" and "too specialized, too localized." He does find the concept of an "etheric (or energy) body" congruent with his theories, though. "Morphic fields can be of many types, and a morphogenetic field is one that organizes and, to a degree, animates bodies," he explains.

The idea that inheritance doesn't come through genes alone is one that carries over into the aging of cells. Although his research on cellular aging, culminating in an article in *Nature* magazine, was done before he formulated his current theory, Sheldrake finds it applicable. "A lot of information is inherited through morphic resonance rather than through genes," he says. "Since these fields contain an inherent memory, they can evolve and change." A vegetarian for twenty-five years (mostly for ethical reasons), he feels we can influence the aging process through diet, exercise, and meditation, but cites the accumulation of defects in cells as ultimately irreversible.

"We can't reprogram our cells completely; though we can slow the aging process, I don't think we can completely reverse or stop it. Aging is a mechanistic process that works against morphic fields." Realistic and practical, Sheldrake is somewhat understated, but can deliver a punch when necessary. He muses that "there's a lot we already know, but we've been educated to reject our own experience. I think it's better to pay more attention to what we see in our animals and experience ourselves and not be frightened off by the dogmatic, mechanistic, and materialistic attitudes that still prevail in the scientific and medical worlds."

Currently a Fellow of the Institute of Noetic Sciences in San Francisco, Sheldrake sees one effect his "fields" have as "influencing personal responsibility and intentions." Noting that social fields can build upon the energy within and around them, contributing to group actions such as "mob violence," he cautions that "morphic resonance is morally neutral. We need to realize that our thoughts and intentions are influential and take responsibility for how things are evolving on the planet. Right now the biggest spreading habit in the world is growing consumerism . . . kids everywhere want to mimic kids in the United States. But think what could happen through prayer and meditation groups!"

For Sheldrake, things are evolving very nicely, thank you. More books are planned, including another with theologian Matthew Fox, with whom he wrote *Natural Grace: Dialogues on Science and Spirituality* and *The Physics of Angels* (both 1996), and one detailing the results of his experiments. He is content to be living as balanced a life as one can in this hectic world, and is happy with his role of being in the vanguard of scientific research and discovery.

33 Telephone Telepathy

Skeptical Outrage Notwithstanding, Rupert Sheldrake Seems to Be Winning the Argument for ESP

John Kettler

The Holy Grail of psi research has long been a simple, readily replicable experiment that yielded statistically significant results, preferably way beyond mere chance. Guess what? Iconoclast biologist, scientific freethinker, and author Dr. Rupert Sheldrake, perhaps best known for his morphogenetic theory—which holds that matter is a kind of flesh applied over an existing energy skeleton in all life (as perhaps exquisitely exemplified in the famous Kirlian photograph of a cut leaf)—appears to have done just that, pioneering large-scale psi research using modern-day household and personal electronics in the process.

In an elegantly simple experiment using household phones without caller ID and a sample pool of four people, all emotionally close to the person (receiver) who must say before answering the call who is calling, where statistical expectation on blind guessing would be 25 percent, Dr. Sheldrake is consistently getting, based on hundreds of trials, 45 percent correct answers among his receivers, nearly twice as good as random chance. This is astounding by both psi research and standard scientific standards, so much so that Dr. Sheldrake published an invited paper, "Testing for Telepathy in Connection with E-mails," in the peer-reviewed scientific journal *Perceptual and Motor Skills,** after publishing a series of papers on telephone telepathy in parapsychology journals, such as the *Journal of Parapsychology* and the *Journal of the Society for Psychical Research.*†

With full experiment duration video coverage at both ends, to prevent cheating via any number of means, these results have not only held up but also have been duplicated by the University of Amsterdam.

---

*Rupert Sheldrake and Pamela Smart, "Testing for Telepathy in Connection with E-mails," *Perceptual and Motor Skills* 101, 771–86.

†For a complete list of his readable online and/or downloadable scientific papers, please see www.sheldrake.org/Articles&Papers/papers/telepathy/index.html.

Contrast this with the old-school techniques now known to be excellent psi killers, what with glowering, intimidating scientists in white lab coats and sterile labs, boring manual, and later mechanized and then computerized die rolling and coin flipping, staring at those Zener cards (square, circle, star, and so on), and so forth. Even worse were the Ganzfeld experiments in which the receiver was in sensory deprivation with eyelids covered first in cotton, then in blacked-out half Ping-Pong balls, listening to piped-in white noise on headphones and bathed in red light while trying to determine which image based on randomly selected video was being telepathically sent.

Fig. 33.1. Experiments with telephones have been used by researchers such as Rupert Sheldrake to investigatge telepathic phenomena. (Illustration by Randy Haragan for Atlantis Rising)

The "observer effect" in psi research is now well known. Those researchers who are open to psi or favor it report results better than chance. Those agnostic on it run basically at chance. Those who don't believe in it or are actively hostile consistently get results well below even chance. The observer effect has long been the bane of truly effective psi research.

THEN ALONG CAME SHELDRAKE

Armed with that rarest of commodities in psi research, real university funding in the form of the top source in Britain, a Perrot-Warrick Scholarship (chartered as "absolutely for psychical research") from the prestigious Trinity College, Cambridge, Dr. Sheldrake essentially took the entire previous approach to psi research and threw it out the window, replacing it with a simple approach that took the research out of psi-killing lab environments and put it into the place where most people feel safe and relaxed, their homes, where they are therefore best able to perform psychically. As study after study has shown, telepathy is most often found between people who are close. Two well-recognized examples have even found their way into TV and film: the link between twins and the link between either a wife and her husband or a mother and her child, famously seen in any number of war movies in which the mother or wife wakes up screaming in the night knowing that her dear one has been killed days before the dreaded War Department telegram arrives.

Further, he almost single-handedly moved psi research to the telephone, e-mail, and text messaging, not only taking unprecedented advantage of burgeoning modern technology, but by doing so also tapping directly into that pool of boundless energy and unceasing curiosity—teenagers—thus drastically expanding his experimental base.

He did this to such phenomenal effect that on September 7, 2006, he appeared as part of a live audience panel discussion for BBC's *Radio 4* program "The Material World," broadcast from the British Association for the Advancement of Science's Festival of Science at the University of East Anglia in Norwich, U.K. The audience was polled at the beginning of the show to determine its relative acceptance of matters psi and at the end to determine what change in views, if any, had occurred. As the host Quentin Cooper was at pains to point out, the B.A.A.S., widely perceived as a citadel of traditional science, has a long, carefully suppressed history of distinguished members with "out there" views, including physicist and chemist Sir William Crookes, who discovered thallium and helium but famously attested that medium David Douglas Home levitated out a window several stories up, floated around the corner of the building, then came in through another window, all while under direct observation by the eminent scientist. These days, we think of him mostly because of the Crookes radiometer, that light-driven whirligig you can buy at science museum gift shops.

The online article on that show is quite good (www.bbc.co.uk/radio4/science/thematerialworld_20060907.shtml) but the half-hour broadcast itself (hit the "Listen Again" link, once there) is remarkable in more ways than we can even list, bringing together a variety of scientific backgrounds and views, some skeptical, in a wide-open, respectful discussion not just of the experiments and proper methodologies, but also of fundamental issues, scientific philosophies, the difference between the true "show-me" skeptic and the false "skeptic" who is really a "no-evidence-will-ever-convince-me" scientific fundamentalist in skeptic's clothing; also discussed were how to guarantee bad results in psi research and the vexed dilemmas of the scientific tendency to not want to publish marginal results, peer reviewed journals generally not being interested in publishing such experimental outcomes, difficulty of getting psi articles published in mainstream scientific journals to begin with, and that great bugaboo of academia, obtaining funding.

There was a droll suggestion offered that those conducting remote influencing experiments should seriously consider hanging around outside university funding committee meetings! The end-of-show audience poll found that several psi contrarians had become psi believers, with none going in the other direction.

"The Material World" was by no means the only such appearance by Dr. Sheldrake. Indeed, BBC *Radio 5 Live* ran an interview of him conducted by skeptic Professor Atkins, but the link (www.sheldrake.org) is regrettably corrupt and works only briefly. A pity, really, seeing as how the Oxford University chemist was quoted as saying, "Although it is politically incorrect to dismiss ideas out of hand, in this case there is absolutely no reason to suppose that telepathy is anything more than a charlatan's fantasy." Nor was his the only scientific establishment voice raised in howling dissent, as ably captured that day in the *Times* (www.sheldrake.org/D&C/controversies/Times_report.html).

"British Association for the Advancement of Science. Theories of telepathy and afterlife cause uproar at top science forum," the *Times*' science editor Mark Henderson wrote. "Scientists claiming to have evidence of life after death and the powers of telepathy triggered a furious row at Britain's premier science festival yesterday. Organizers of the British Association for the Advancement of Science (the BA [shorthand for the B.A.A.S.]) were accused of lending credibility to maverick theories on the paranormal by allowing the highly controversial research to be aired unchallenged.

"Leading members of the science establishment criticized the BA's decision to showcase papers purporting to demonstrate telepathy and the survival of human consciousness after someone dies. They said that such ideas, which are widely rejected by experts, had no place in the festival without challenge from skeptics."

No doubt Professor Chris French, a skeptic from London's Goldsmiths College who sat on that previously described panel and was interviewed live, would take exception to such an assertion!

Here are a few more samples, as cited in the above article. Lord Winston, fertility specialist and former president of the British Association for the Advancement of Science: "I know of no serious, properly done studies which make me feel that this is anything other than nonsense. It is perfectly reasonable to have a session like this, but it should be robustly challenged by scientists who work in accredited psychological fields."

Sir Walter Bodmer, geneticist and president of Hertford College, Oxford: "I'm amazed that the BA has allowed it to happen in this way. You have got to be careful not to suppress ideas, even if they are beyond the pale, but it's quite inappropriate to have a session like that without putting forward a more convincing view. It's extremely important in cases like this, especially for the BA, which represents science and which people expect to believe, to provide a proper counter-argument."

Readers are free to decipher the unmistakable meta-messages at their leisure, but what's really being said by these worthies is pretty obvious, to this writer at least. With such a fabulous target, the *Times* editorial staff had a field day, assailing not merely the scientific establishment but also a whole bunch of secrecy-driven others in most withering terms (www.sheldrake .org). It's short, pithy, and utterly merciless in its conclusions. As such, given its conclusions and the highly respected, influential periodical in which it appears, it may eventually be seen as the first mainstream death knell for the old rationalist/reductionist/materialist model of reality.

BRAINSTORM

It goes without saying that telepathy works. Are you thinking what we're thinking? We thought so, and ordinarily that would obviate the need for this article.

But it seems that some skeptics are still clinging—deaf to the inner voices of their own minds—to the dull empiricism that deems telepathy not just implausible but rather impossible. For them, it is worth spelling out: The evidence for mind-to-mind communication is now so overwhelming that it has been presented at the annual meeting of the British Association for the Advancement of Science. This has dire implications for the telecoms and polygraph industries. The treasured careers of secret agents, diplomats, and war reporters will suffer if they cannot master this newly respectable skill. Professional poker is doomed, lying is more perilous than ever, and human courtship will never be the same again.

Full disclosure requires that we acknowledge in old-fashioned print, as well as via what Upton Sinclair called "mental radio," the depth of dismay occasioned by Rupert Sheldrake's presentation at the B.A.A.S. of the results of the experiment in which participants were invited to guess which of four friends was about to call them on the telephone: correct answers 45 percent of the time, proving the existence of telepathy. Crusty traditionalists say that this proves the existence of nothing but intuition and dumb luck. We say, ponder the *Times'* input-output theory, a.k.a. "Tinpot": "We take in so much food, water, and inspiration, and our cerebral cortices generate so much measurable electrical activity that it is absurd to think that none of it escapes in the form of transmittable brainwaves. That's what we're thinking, and it's the thought that counts" (the *Times* editorial, September 6, 2006).

One link that does work, though, is Dr. Sheldrake's October 5, 2006, appearance on ABC Radio National's *In Conversation,* hosted by Robyn Williams, where readers may not only listen to the audio but also read the transcript. The show, though, is concerned less with the experiments themselves than with how Dr. Sheldrake, once a "very respected botanist," got so scientifically far afield. It is an inquiry into who he is these days, how he sees himself, and what he thinks. In the interview he reports that some fifteen hundred trials involving telephone telepathy and telepathy via e-mail have been done, and that the odds against nearly three hundred telephone telepathy trials yielding 45 percent accuracy by random chance stand at around a billion to one (www.abc.net.au/rn/inconversation/stories/2006/1754367.htm#).

DON'T TAKE OUR WORD FOR IT

Those of you who like to experiment for yourselves may want to go to his Experiment Portal, where you can learn how to participate in a variety of large-scale trials spread out all over the world (www.sheldrake.org/Online exp/portal/).

How spread out? His tests found that telepathy was completely range-independent, as verified in tests originating in England and received in Australia. Interestingly, earth sensitives, covered in "The Bio-Sensitive Factor" (*Atlantis Rising* 27) by this writer, have found the same thing to be true, for they can feel the buildup of crustal strain and volcanic pressure clear on the far side of the planet, as well as the earth events themselves when they strike. Some can even detect solar upheavals before the actual emissions reach the scientific monitoring satellites.

Of course, this neatly fits into the emerging notion of the interconnectedness of everything, whether expressed scientifically in the exotic terms of quantum entanglement, as the in-forming principle of metaphysics, the Manitou of the Native Americans, the "Force" familiar to legions of *Star Wars* fans, or Dr. Rupert Sheldrake's theories of morphic resonance and the morphogenetic field. We should note, too, his studies of the connection between intentionality and the various telepathic phenomena he studies, as well as the fact that, using a special SQUID (Superconducting Quantum Interference Detector) helmet, Viennese experimenters succeeded over a decade ago in scientifically detecting and measuring the intention signal generated by the human brain before any electrochemical "orders" were sent to the muscles via the nervous system (200 mph). Per *Leading Edge*, ". . . micromagnetic fields . . . preceded voluntary activity by 30–50 milliseconds. . . ." The implication is shattering—"intent is a luminal process that bypasses the ordinary neural networks of the brain."*

Is it this electromagnetic energy that a telephone telepath reads and analyzes when a call comes in, so as to determine who the caller is before the caller so much as breathes?

Is it this that a sentry—as a potential victim with back turned—picks up and responds to via an abrupt about-face, having read the stalker's intent to kill before the stalker even moves?

When we are in great peril or about to die, does our cry for help go forth on literal wings of light, coming to rest with our beloved?

These are but some of the key questions Dr. Rupert Sheldrake is investigating these days, questions that, whether or not he intends it, go to the very essence of who we truly are, what we're about, and why we're here.

---

*Leading Edge 78 (January 1995): 24.

PART EIGHT

THE ET FACTOR

34 Back-Engineering Roswell

An American Computer Maker Fights for Its Right to Exploit Alien Technology

John Kettler

After asking a question on its website someone didn't want answered, a small company is fighting for survival and the people's right to know the truth. The company is American Computer Company (ACC), located in Cranford, New Jersey, under the leadership of computer design pioneer Jack Shulman. The question, posed as a humorous "what if," asked whether the transistor was perhaps not invented at Bell Labs, but was instead discovered in the wreckage of the so-called Roswell crash, then reverse engineered.

Shulman and his company believe they have evidence—primarily a massive "Lab Shopkeeper's Notebook"—that such reverse engineering occurred. Acquired from anonymous sources, the "notebook" purports to describe in detail the entire engineering process by which Bell Labs developed the technology taken from the Roswell craft in 1947 into a number of important patents now held by Bell and other major players in the aerospace and computer technology industries. The notebook's authenticity, ACC says, has been independently verified by forensic examination. The company also believes it has identified major technologies in the material that are yet to be developed, and it wants the right to exploit them on its own.

Fig. 34.1. American Computer Company president Jack Shulman.

What does it mean if ACC can show that Bell Labs didn't invent the transistor, if the U.S. government simply found a wreck full of exotic technologies, figured out how some of them worked, then secretly and illegally gave priceless technology to Bell Labs? What if Bell Labs then reverse engineered the technology, patented it as its own, produced it, and made billions from it?

What does it mean if patents were fraudulently awarded, undeserved technical reputations were acquired, and Nobel Prizes were granted for non-existent scientific breakthroughs? And what of the Roswell crash itself?

The official version of events is that no alien craft crashed to earth. Yet there is mounting evidence that something (origin unknown) did, and that it was recovered and exploited for its incredibly advanced technology.

This scenario has been reinforced by the independent testimony of retired Army Colonel Philip Corso in his stunning book *The Day After Roswell*. In it, Corso explicitly describes being tasked in 1961 by the then head of Army research and development, General Trudeau, to untraceably transfer to Bell Labs, IBM, Monsanto, and others, exotic technologies and devices acquired by the Army from the Roswell crash. The companies were to develop and patent the technologies and devices as their own.

These are precisely the issues that ACC believes have generated many powerful attacks on it. The company (www.accpc.com) makes PCs, Web servers, and small supercomputers under what CEO and President Shulman calls a "virtualized corporate structure." Simply put, this means a small cadre of highly skilled people does the critical design and engineering work, but many other tasks are contracted out.

Attacks, say company officials, following its Roswell question have been massive. They range from mindless Internet rants to life-threatening acts. Also included were a break-in at company headquarters, the planting of classified material on the premises, and the bizarre receipt of a classified fax pertaining to a previously unknown "deep black" space platform.

One major theme of the attacks is that ACC either doesn't exist or is so small and inconsequential that no one should pay attention to it or its shocking discoveries. The former can be summarily disposed of, since the firm is listed with Dunn and Bradstreet and is licensed to do business in New Jersey. Another example is an Air Force claim stating ACC is but a "one room office in a busy business complex." The firm actually occupies

Fig. 34.2. Retired Army colonel Philip Corso, whose book The Day After Roswell *recounts his participation in the Roswell cover-up, testifies before Congress.*

a suite of offices at its headquarters and also has facilities elsewhere.

Another method of attack is postings to sites all over the globe under a variety of "handles," ranging from the annoying to anti-Semitic slurs at Shulman himself. The evident goal is to smear and discredit ACC and Shulman in the eyes of the public.

The company has responded aggressively, initiating comprehensive surveillance and intrusion countermeasures on its many sites to detect, defeat, and backtrack to the source of efforts to hack its sites.

Carefully monitoring what is being said about it on the Internet, ACC reacts swiftly to slander and libel wherever it finds them. One site received such emphatic attention from ACC's legal counsel that it subsequently posted a request for peace to ACC's "Alien Science and Technology Forum."

ACC has publicly alleged that personnel working for Lucent Technologies, the U.S. Air Force, and Wackenhut Security have been involved in one way or another in such attacks. Full details are available on ACC's Roswell site and in the hyperlinked Alien Science and Technology Forum.

On March 18, 1998, in Union County Superior Court, New Jersey, the company filed a slander and libel suit (Docket #UNN-L-1751-98) against Lan Lamphere and Windchaser, Inc. Mr. Lamphere and Windchaser are alleged in the suit not only to have libeled and slandered ACC and Jack Shulman, but also to have made veiled threats and to have acted in such a way as to encourage others to commit violence against Mr. Shulman, key ACC personnel, and their families. The defendants are attempting to have the suit vacated.

ACC is fighting more than perceived libel, though. The company says it has another lawsuit primed and ready to file, against the Air Force, with respect to information disclosure regarding the Roswell crash. If it files, the company intends to attempt to use the Air Force's own regulations against it. ACC believes that the Air Force may have violated regulations by "parking" a Roswell craft under one of its classified research aircraft programs, a project named Silverbug, under which a series of turbojet-powered, saucer-shaped aircraft were developed and presumably test flown.

An air show attendee at Nellis AFB some years ago may have seen Silverbug after making a wrong turn in his Jeep, which was the same dark blue as those driven by the Air Police. Thus camouflaged, he apparently entered a sensitive area undetected, in time to witness a silvery metallic disk—complete with bubble canopy and pilot in full Air Force flight garb at the controls—blast off with a roar.

The drawing below, obtained by ACC from the Air Force by a Freedom of Information Act (FOIA) request, certainly appears to show something like what the witness described. Unfortunately, the Air Force has yet to release this unusual aircraft's performance figures. What can be said, though, is that the design appears to draw on both the German saucer designs (previously covered in issue 7 of *Atlantis Rising*) and the derivative Avro saucer depicted in the June 1956 issue of *Look* magazine.

The suit, if filed, will demand that the Air Force publicly produce said Roswell craft, its crew remains, and the analyses done of the craft and crew—if such exist. If the Air Force claims, "We don't have them; it never happened," then, ACC believes, the service will be legally required to provide a rigorous and exhaustive explanation for the gigantic program of suppression, leaks, disinformation, documents, and so on originating from within itself and elsewhere in the government since 1947 and continuing today. Of necessity, this includes any actions taken against the citizenry in connection with this program.

In the meantime, before filing suit, ACC is working through its administrative options. As Jack Shulman noted, "Government judges love to throw out suits if we don't first exhaust all avenues under the Freedom of

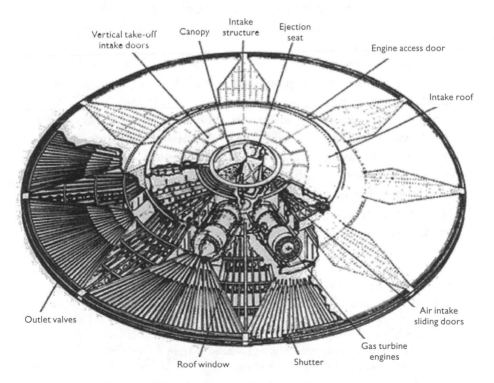

Fig. 34.3. Technical illustration of the Silverbug craft.

Information Act."* He also remarked that Dr. Peter Gersten and CAUS (Citizens Against UFO Secrecy) filed their landmark Roswell lawsuit against the Army and the Air Force while ACC was getting no response from the Air Force to its queries filed via public means of discovery.

ACC has never said that the Roswell craft was alien—in spite of assertions to the contrary. It has simply said that the craft was not of domestic origin. While indicating that it might be alien, ACC has also explicitly held open the possibilities that the craft may also have derived from World War II German or Japanese "black" programs, from scientific slave labor or from émigré scientists who fled Germany before the war.

Bill McDonald, a forensic investigator whose reconstruction of the Roswell craft has become the basis of a popular model kit, has no such doubts. Based on rigorous interviews of surviving eyewitnesses, including Counter Intelligence Corps men who were there, and after poring over reams of testimony compiled during five years of investigation, McDonald says he knows the Roswell spacecraft was created by an extraterrestrial civilization, a society that lost five of its members in coming here to investigate our atomic tests, which he likened to "a cosmic smoke signal." In describing both the Roswell spacecraft and its flight crew, he speaks in hushed, almost reverent tones. "Alien" is not a term he uses. "It's not respectful," he said.

ACC is now ready to move forward with the launch of several new technologies that it has discovered in the thousand-plus-page Lab Shopkeeper's Notebook, which, ACC claims, have been "overlooked since 1947."

Three discoveries have so far been publicly identified. They are: the transfer capacitor, being trademarked by ACC under the names TransCap and TCAP; the photonitron; and magnetic liquid memory.

The TCAP is apparently a kind of electronic nerve cell from an almost sentient neural computer array allegedly used to operate a Roswell craft. ACC has already arranged to mass-produce a field trial version of the device, which is used in conjunction with a digital power manager to stretch battery life some 13 percent on special editions of its lately produced Tiger and

---

*In order to sue the Air Force in the U.S. Court of Claims, which has jurisdiction over it, the party bringing suit must first exhaust all available legal remedies (every potential solution short of actual trial, to include binding arbitration, if mutually agreeable). Given the government's essentially unlimited supply of lawyers, and lawyers' proven skill in delaying the progress of lawsuits to trial, it could take twenty years, assuming the case isn't resolved meanwhile, before ACC's case would even be eligible to be heard.

Pantera notebook PCs. This may not seem like much, but notebook PC battery life is typically only a few hours. The TCAP works well in such applications because its voltage requirements are tiny and because it is nearly lossless, something like a room-temperature semiconductor.

ACC has also produced and marketed a series of proprietary TCAP design and application kits, targeted on everything from small specialist firms to industrial giants. Readers wanting more information should go to ACC's Roswell website. ACC won't sell the technology, but licenses it for specific purposes and end uses to different companies.

Why all the fuss? What kind of commercial advantages accrued to Bell Labs from supposedly inventing the transistor? What if a company had new memory technology—technology 10,000 to 100,000 times faster than what our fastest PCs have now? Or what about solid-state storage devices capable of holding 90 gigabytes (90 billion bytes) of data in the space taken up by a poker chip? That's equal to fifteen full-length movies at laser disk resolution or better.

This technology is in its infancy, but it has already generated tremendous reactions, including an offer, ACC has reported, to buy the TCAP and other technologies for $50 million on condition that it also destroy the Lab Shopkeeper's Notebook. The company passed, even though the money would have been very useful.

ACC has had to deal with many challenges, from hordes of would-be buyers to major firms and the government raising hell, claiming that the TCAP and other goodies are theirs. ACC's reply: "You've had since 1947 to do something about this. Go away!"

The photonitron is a device that Shulman describes as having "both a photonic and a microwave emission." It works somewhat along the lines of a laser hologram projector except that the resultant image is not only three dimensional, but also radar reflective. The military applications of such a device for countermeasures and psychological warfare are tremendous. Shulman says he has evidence that Wright Patterson AFB has a photonitron and is actively testing it. Our query to the base's media office about this went unanswered.

Magnetic liquid memory allows for previously unheard-of information storage densities. The most recent development in the engrossing saga of the TCAP is that ACC has figured out how this device operates, an absolutely essential prelude to filing a patent. Shulman told *Atlantis Rising* exclusively, "We know how it works." ACC had previously announced, on August 4, 1997, that it had completed research and development on two TCAP Proto-Hyper-Storage Devices. PHS I is capable of mimicking an 8.4 GB hard drive; PHS II can mimic a 90 GB hard drive!

He then listed the reasons for proceeding with the filing of a TCAP patent—well beyond the mere protection of intellectual property:

1. "To see whether the U.S. government exercises the National Security component of the Patent and Trademark Law (35 USC Sec. 181 1,6,97)." This section allows the government to arbitrarily classify an invention, regardless of the intended purpose of the inventor or the invention.
2. To establish the basis for legitimately transferring such technologies.
3. "In order to protect it while we go through the EPA Energy Star process." All new computers are required to undergo government certification for energy efficiency. Gaining such certification requires providing the government with highly detailed specifics on the computer's design. The problem? ACC says it has discovered through its Roswell investigation that AT&T, a gigantic and powerful rival, has, via its relationship with Lockheed Martin—the government contractor for the Energy Star program—apparently managed to position itself as a key player in the Energy Star certification process. AT&T reportedly also has a similar indirect "feed" into the Patent and Trademark Office.

Just before this chapter was first printed as an article, ACC announced it intended to proceed directly to production without a patent, and plans to prevent anyone else from patenting the technology by placing all TCAP how-to documents in the public domain.

Long ago ACC's investigation passed 10,000 interviews and more than 100,000 pages of documents. It continues to this day, reinforced, not deterred, by attacks upon it from government and industry alike.

What does the future hold for the embattled company? Can this small firm survive against the power of major corporations and the government?

Sitting at ground zero is Jack Shulman, who makes no bones that he expects things to get really ugly soon. This has everything to do with ACC's drive to secure patents and take its revolutionary new technologies to market, threatening vast business empires and billions in profits.

Is he worried? Perhaps. Will fear stop him? "I don't scare easily," he says.

35 The Fight for Alien Technology

Jack Shulman Remains Undaunted by Mounting Threat

John Kettler

Issue 16 (1999) of *Atlantis Rising* presented a story that read like something right out of *The X-Files*. Titled "Back-Engineering Roswell" (chapter 34 in this book), the article details the story of a small New Jersey computer firm, American Computer Company (ACC), whose researches and breakthroughs so fundamentally threatened corporate, scientific, government, and intelligence interests that the company was broken into; its networks were hacked from government facilities; a service provider, Softnet, was hacked; and the storage arrays holding ACC's site data, along with dozens of other customers, were destroyed. ACC and its customers were the target of every dirty trick imaginable. ACC's president became the target of a million-dollar extortion attempt, the Mafia was hired to threaten the company, and death threats were made against ACC's key executives. Photos and addresses of the executives' children were even posted on the Internet, along with their schedules, an open invitation to kidnap the children, and prospective customers of the company were contacted and threatened. Deadly serious stuff.

What could cause such a concerted vicious response, and why the ET angle so characteristic of *The X-Files* show? Because there may well be one. ACC was temporarily given access to something called The Lab Shopkeeper's Notebook by the widow of a gentleman named Jeff Proskauer. The notebook was scrutinized by ACC and a bevy of specialists. ACC also launched a major investigation, upon which nearly a million dollars was spent. The objective was to independently confirm or disprove the notebook and the amazing story behind it. What was the story of the notebook? It was allegedly the notes concerning nothing less than the joint 1947 technical exploitation of the famous Roswell craft by Bell Labs and IBM.

Very little of the notebook's contents has been released, but what has been has shocked certain sectors to the core. One discovery was that the transistor, historically considered to have been invented at Bell Labs by Bardeen, Brattain, and Shockley, was described in the notebook as having been "provided" to Bell Labs by the government for reverse engineering. That invention was worth Nobel Prizes, enormous scientific and corporate reputations, a stack of patents, and trillions of dollars. Unfortunately for all concerned, the most searching investigation into the transistor's origins found that the devices that were supposedly the transistor's ancestors utilized a fundamentally different technical approach that would never have led to the transistor.

Consider the implications of that for a few minutes. One of the premier and basic inventions of modern technical civilization may not be ours. It may be a covert ET technology transfer, exactly as described by then retired, since deceased Army Lieutenant Colonel Philip Corso in his revelatory book, *The Day After Roswell.*

He says that he was tasked by his then Army boss, General Trudeau, head of Army Research and Development, to untraceably "seed" the Army's share of the Roswell craft technology treasures, things like advanced night vision devices, super tensile fibers (Kevlar), and fiber optics, to the appropriate companies, with the specific understanding that they would patent and produce them as their own inventions or scientific discoveries. A similar scenario was reportedly described in the since withdrawn from publication autobiogra-

Fig. 35.1. Bell Labs engineers Bardeen, Brattain, and Shockley are credited with inventing the transistor.

phy of IBM World Trade Chairman Jerry Hartsell; there the technology was electronic. IBM World Trade is the little-known foreign distribution arm of Big Blue. It is the part of IBM that deals directly with foreign governments and multinational corporations.

If the above allegations are true, then the entire history of modern science and technology not only may be fundamentally flawed, but may constitute one of

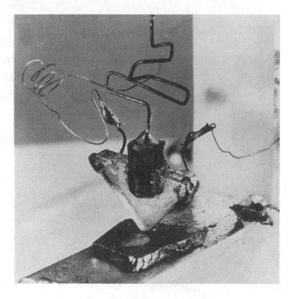

Fig. 35.2. The first transistor, invented in 1947.

the greatest frauds ever perpetrated as well. Could our towering scientific geniuses be little better than the scientific hacks of the former Soviet Union, people who received Stalin prizes for reverse engineering Western technology, technology the Soviets couldn't themselves create?

That shocking disclosure was merely the opening discussion. The notebook revealed another device that took matters into another realm, a realm in which the word *breakthrough* became utterly, profoundly inadequate. Welcome, then, to the world of the transfer capacitor, the TCAP, which has caused a tempest that roars to this day.

A fast desktop PC operates at a clock speed of 1.2 GHz, or 1.2 billion cycles per second. That sounds impressive when you realize that this article is being written on a computer that clocks at a piddly 233 MHz, or 233 million cycles per second. How impressive will that powerful PC be, though, if someone shows up with one that runs at, say, 12 THz, or 12 trillion cycles per second? What if it runs at 50 THz? It's been done. ACC built and ran a breadboard four-function programmable 50 THz calculator. The Lawrence Labs at Berkeley, California, have gotten a much cruder version of the device to run.

The TCAP has other features that add to its attractiveness. For one, even in its present early development state it allows incredible information storage densities, 90 GB (90 gigabytes, 90 billion bytes), the equivalent of

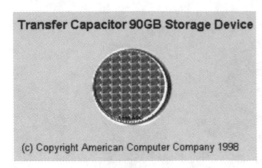

Transfer Capacitor 90GB Storage Device

(c) Copyright American Computer Company 1998

Fig. 35.3. American Computer Company's advanced TCAP Wafer. (Image courtesy of the American Computer Company)

fifteen full-length movies, in the space occupied by a poker chip. Another is that the TCAP is practically a room-temperature superconductor having almost zero energy loss. Its power needs are vanishingly small, wonderful news in an era of power shortages and proliferating electronics. ACC put an early TCAP-based power controller into some of its Tiger notebook computers and got a 13 percent increase in battery life. That may not sound like much until you remember that notebook battery life then was only a couple of hours.

In the previous chapter we mentioned that ACC had decided to proceed directly to production. Here, ACC pulled another surprise on its rivals. The firm, citing a 1967 UN Treaty on Space, which grants the fruits of space exploration to all nations, not the privileged few that can afford to go to space, elected to put the essential technology and theory of the TCAP, the "art" as defined in patent law, into the public domain, thus preventing anyone from ever patenting it. ACC's legal theory to justify this was that its investigation, which cost so much and ran to over 10,000 interviews, showed that the original source of the TCAP was non-terrestrial, and hence was covered under the provisions of the treaty. We know of no instance in which a commercial firm ever voluntarily renounced the possibility of patenting an invention of incalculable ultimate value (the transistor was worth trillions) and made it available to all of humanity.

TCAP technology hasn't remained dormant. ACC and its research partners have drastically redefined the upper limit for device performance and lowered the limit for size. The best available estimate for TCAP clock speed is now an incomprehensible "2 million gigahertz." Device size is a function of what must happen internally in order to make the device work. Since a TCAP works by temporarily shifting the orbital state of a single electron during one fraction of an orbit, it logically follows that this wouldn't take much space or time. It doesn't. The best estimate is that device size can be submolecular. Yes, you read that right: several devices per molecule. As a storage device, the TCAP's potential is equally mind-warping—it's a function of the numbers of atoms in the storage matrix and their arrangement. Furthermore, ACC recently revealed that the TCAP could be used to manipulate light. Therefore, reflected light beams can be used to read data directly from TCAP memory via fiber-optic lines, leading to direct optical communications between computers at the incomprehensible speeds of the TCAP. This newer variation of the TCAP is known as the TCAP Optical Interferon.

All such progress has been dearly purchased, for ACC was and is a company beset yet ever game, a tale chillingly told on Comp America Online's "Alien Science and Technology Forum" (www.aliensci.com) in the form of libelous postings to the forum designed to confuse people and sow dissension against ACC, in the string of stories about hackers and the damage (intrusions, denials of service, wrecked routers and switchers, whole networks crashed) they've caused or tried to cause ACC and those who supply it with Internet services and equipment, in frauds and thefts committed against ACC, and even in what happens to those who post on the boards. Then there's the current economy and its impact on the computer business.

As anyone who follows Wall Street or information technology knows, the early 2000s saw a major retrenchment in the computer business, and sales dropped so dramatically that many of the majors took huge stock price hits. Those hits, in turn, translated into large staff cuts.

Such times are worse for a small firm, even a privately owned one like ACC, for a small firm simply doesn't have at its disposal the immense resources or the credit lines of its giant rivals. It really hurt when ACC, an online business with many websites, was suddenly unplugged after its DSL provider, Northpoint Communications, was forced into bankruptcy and its assets bought for pennies on the dollar—by AT&T, the successor to Bell Labs and one of ACC's fiercest opponents in the TCAP wars. (AT&T claims the TCAP as its own.) This action not only completely disrupted normal business for ACC in New Jersey, but it digitally orphaned 600,000 Northpoint DSL subscribers in California. That led to an official complaint to the Public Utilities Commission (PUC) of California, and the PUC ruled that AT&T could not arbitrarily cut off service and convert the lines to its own use. That was the good news. The bad news was that though the PUC was absolutely right in its ruling, because this coup de main was done under bankruptcy law, the PUC was powerless to apply sanctions, as Northpoint no longer existed and therefore there was no one left to be compelled to continue provision of services by the PUC.

The forced disconnect disrupted ACC, but only briefly. A new threat has emerged, the New Jersey Mafia, allegedly in the living counterparts of the HBO hit series *The Sopranos,* an ongoing crime saga set in New Jersey that details the lives of a Mafia underboss and his crews.

The "garbage business" may be Tony Soprano's cover story, but Anthony Rotundo, of the DeCavalcante family, who was revealed to be quite a fan of the show by FBI undercover tapes released at his bail hearing—"What characters! Great acting!"—apparently was less successful with his cover, being named in a sweeping federal RICO (Racketeering Influenced Corrupt Organization) Act indictment in late 1999 that charged him and forty of his associates with murder, conspiracy, extortion, and gambling.

ACC alleges the following crimes against it by the North Jersey Mafia: extortion, financial scam after scam, receipt of bogus checks ranging from $25,000 to $50,000, large-scale theft of money and services ranging from $25,000 to $120,000, and large-scale theft of computers by their front companies by simply not paying for them. This picture was expanded somewhat via "deep background" interviews.

When contacted and asked directly by this author whether ACC had filed criminal or civil complaints, Jack Shulman, ACC's chief technologist and CEO replied, "I'm not going to answer that question." He did indicate that it was best not to comment about ongoing litigation, collection matters, and criminal investigations once they were under way.

Jack Shulman also issued a wake-up call to those who think the Mafia's day is done. "The Mafia isn't dead; it's just gotten a lot smarter. Today, they get jobs providing protection to, well, IBM and Merrill Lynch. Yesterday, it was just garbage, trucking, and entertainment!"

This is exactly what our digging confirmed. The Mafia's gone high tech, following the money into Wall Street stock swindles, the Internet (porn, gambling, online fraud, and so on), banking, gun running and illegal technology transfer, people smuggling, and much more, in addition to its traditional crimes.

The new Mafia is characterized by a low-profile corporate approach and interlocking global networks and connections, the same structure that terrified the authorities when they began to understand the true nature of international terrorism. This time, it's much worse. The new Mafia finances itself, buys and sells officials, and operates from safe havens worldwide.

In an article entitled "New Face of the Mafia in Sicily. High-tech transformation—with global tentacles" originally written for the *San Francisco Chronicle* on January 8, 2001, and reprinted online at AmericanMafia .com, staff writer Frank Viviano depicts the emergence of the new Sicilian

Mafia, called in Sicily the Cosa Nuova (the new thing, the new organization). He describes the new Mafia, which rose from the ashes of the shattered old one as being so financially savvy that Sicily's official economy is Third World, yet somehow there is enough money to allow car dealerships on almost every block of some Palermo neighborhoods and let Sicily lead the nation in the consumption of certain food delicacies.

Both more disturbing and more pertinent to Mr. Shulman's point, the Cosa Nuova was barely stopped from taking over the University of Messina. Mr. Viviano writes: "On Oct. 18, hundreds of Italian police in masks and riot gear descended on the University of Messina, one of southern Italy's premier educational institutions. When the smoke cleared, 79 faculty and staff members had been formally indicted on organized crime charges. Their numbers are expected to climb substantially as the case unfolds."

Investigation disclosed that bribery, intimidation, and blackmail brought the arrested into the fold, and that the majority were concentrated in the university's science and technology departments. The article goes on to note that this near hijacking of an entire university, for the authorities were barely in time, was the inadvertent result of good intentions and a bad economy. People who used to have a high school education were now college grads, and they wanted high-tech work in a region largely devoid of it.

It's exciting at ACC, with its super advanced tech, attacks by giant commercial rivals, overt and covert government interference (see the Comp AmericaOnline.com websites; www.aliensci.com and www.roswell.internet.com for details), attempts to defraud the company for work done by it, the Mafia, hackers, a tough market environment, and a blast from the past—the continued assaults on the TCAP and character assassination of ACC execs including Jack Shulman.

"On the off chance that you are a real person and not one of Shulman's aliases, you should look at . . . for info on bizarre Jack and the imaginary TCAP. —Newsboy" (an unsolicited e-mail sent to the writer after he visited the Comp AmericaOnline.com website known as RoswellInternet.com).

PART NINE

OTHER DIMENSIONS

36 Altered States

Recent Research Sheds New Light on the Inner Reaches of Human Consciousness

Patrick Marsolek

You may have seen images of an entranced person lightly walking bare-foot across a bed of red-hot coals without a single blister. Perhaps you've heard about how a hypnotized person sitting quietly in a chair can feel no pain when a needle is stuck into his arm. The person may even have his eyes open and be watching the process. This is the type of phenomenon often associated with trance and other altered states of consciousness.

For several hundred years Western thinkers have distrusted these states. That view may be changing. At present neuroscientists, physicists, psychologists and psychiatrists, medical doctors, and parapsychologists are all trying to understand how these types of phenomena work and whether or not they have value.

An altered state of consciousness (ASC) is generally defined as any mental state that is perceived by an individual, or an observer, as being significantly different from normal, waking consciousness. These ASCs may range from ordinary daydreams to experiences of mystical ecstasy to near-death experiences (see the accompanying graph). A person can tell if he is in an altered state by any of the following signs: alterations in thinking, dis-

Fig. 36.1. An Indian holy man prepares for deep meditation.

turbed time sense, loss of control, changes in emotional expression, changes in body image or sensation, and perceptual distortions.

All ASCs are deviations from our normal consciousness. Charles Tart, who wrote *Altered States of Consciousness* more than thirty years ago, proposed that this normal consciousness should in fact be called the "consensus trance." This is because how we perceive reality is a construct of our

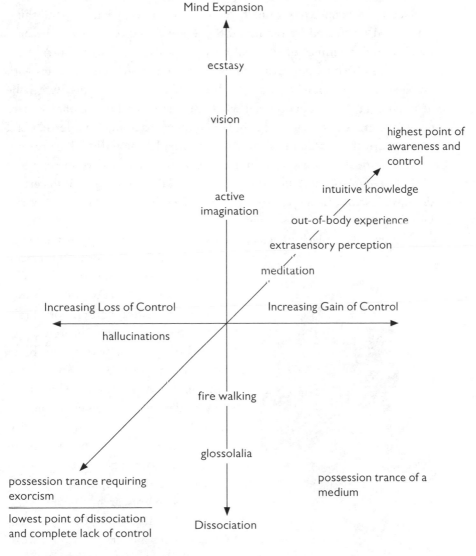

Fig. 36.2. A diagram mapping different types of consciousness. (Graph courtesy of Patrick Marsolek)

beliefs and cultural conditioning. Any time we perceive a belief as absolute or unchangeable we are in trance. The "entranced" way we live may explain why we have so much difficulty understanding trance and ASC. Advocates and skeptics of the value of these states are both firmly entrenched in their beliefs about them. Is there a way to understand these altered states outside of belief? Let's take a look at several avenues of scientific inquiry that are bringing together both what we know objectively about the brain and what we know subjectively from our own experience.

Neurologists have traditionally felt that all we see, hear, feel, and think is mediated or created by the brain. Some are trying to discover the neurological underpinnings of spiritual and mystical experiences. Dr. Andrew Newberg has been mapping the brains of meditators in mystical states with radioactive tracers pumped into the brain at critical moments, which he then photographs. He reports that what really stood out in the photos were the quieter areas of the brain: "A bundle of neurons in the superior parietal lobe, toward the top and back of the brain, had gone dark." This region, called the orientation association area, tells us where we are in time and space. It requires sensory input to function. When it quiets down in certain ASCs, we lose the distinction between ourselves and the world; we perceive everything as self, interwoven and connected.

This activity, or lack of activity, shows how brain function is related to these states. Does it mean the nature of these altered state experiences is mechanical? Not necessarily. Consider if you were to photograph your brain while you were eating an orange. All the neurological activity in the brain wouldn't negate the reality of the orange. Newberg says, "There's no way to determine whether the neurological changes associated with spiritual experience mean that the brain is causing those experiences . . . or is instead perceiving a spiritual reality."

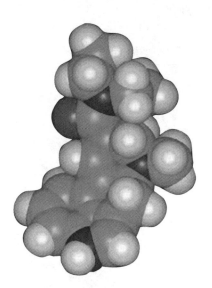

Fig. 36.3. A model of the LSD molecule.

In related research, Michael Persinger, of Laurentian University in Canada, uses a device to send a weak magnetic field into people's heads to influence their temporal lobes. This creates experiences described as mystical, out of body, or even like haunt-ings. In one study a woman's nightly visitations by the "Holy Spirit" were found to be caused by a clock on her bedside table. The "magnetic pulses generated by the clock (were) similar to shapes that evoke electrical seizures in epileptic rats and sensitive humans." In another experiment a journalist who had previously experienced a haunting reported "rushes of fear" and a visual apparition, which he said was very similar to his original experience. Persinger suggests that this type of experiment may help researchers under-stand what environmental variables give rise to the original occurrences of this kind of phenomenon.

In another paper he seemingly proves his point. He correlated experi-ences attributed to Christ and Mary at Marmora, Ontario, Canada, to the location of an open-pit magnetite mine that has been filling with water. He noted that epicenters for local seismic events have also moved closer to the pit. "Most of the messages attributed to spiritual beings by 'sensitive' indi-viduals occurred one or two days after increased global geomagnetic activ-ity," he said. This research clearly seems to offer a causal, non-paranormal explanation for some spiritual experiences.

Some researchers believe that when areas of the brain, like the orienta-tion area, become quiet, it is a regression from higher functioning to a more primitive state, unthinking yet aware. Laurence O. McKinney writes that the state of "selfless perception would be experienced as a state of grace to a religious Westerner, samadhi or satori to a Hindu or a Buddhist." Except he says that this self-induced state is a "lower consciousness in fact." McKin-ney believes that these experiences can be positive, that "moments of mild ego loss are instructive, not destructive, because they were done purpose-fully. . . . Every time we repeat thoughtfully something that we love to do, we add to our growing networks of associative energy." Are these states a regression to a more primitive functioning that is beneficial only because it's managed by the higher consciousness of normal cognitive functioning?

Neuroscientist Rhawn Joseph questions assumptions like this, "Why would the limbic system evolve specialized neurons or neural networks . . . to experience or hallucinate spirits, angels, and the souls of the living and the departed if these entities had no basis in reality? We can hear because there are sounds that can be perceived and because we evolved specialized brain tissue that analyzes this information. First came sounds, and then later, specialized nerve cells evolved that could analyze vibrations and then later, sounds. Likewise, if there were nothing to contemplate visually we would not have evolved eyes or visual cortex, which analyzes this information.

Visual stimuli existed before the neurons that evolved in order to process these signs. Should not the same evolutionary principles apply to the limbic system and religious experience?"

Neurosurgeon Wilder Penfield's research on epileptics significantly increased our understanding of the relationship between the brain and the mind. He discovered that since the brain has no pain receptors, he could directly stimulate the brain of a conscious patient. For example, he would stimulate one spot and the person's arm would move, another spot and the person would suddenly smell lemon. Penfield conducted volumes of experiments showing how various experiences were located in different areas of the brain. Although he found that the content of consciousness depends in large measure on neuronal activity in the brain, this activity "always occurred within the dominating and enveloping radiance of an autonomous mind." His research failed to show where the mind resided in the brain. Later in his career he went so far as to say that although all his experiments were built on the principle that the brain generates the mind, they in fact proved exactly the opposite.

Dr. Les Fehmi, a psychologist and neurofeedback researcher from Princeton, is also studying the value of subjective experience as well as what we know about the physical mechanisms in the brain. He promotes an open focus state of awareness signified by synchronous alpha frequencies in the brain. He first experienced these alpha frequencies for himself when he tried and failed. "At the moment of surrender I experienced a deep and profound feeling of disappointment. Fortunately, I surrendered while still connected to my EEG and while still receiving feedback. It was surprising to observe that I now produced five times the amount of alpha than before the act of surrendering." After learning how to open his focus and create the alpha waves, he "felt more open, lighter, freer, more energetic and spontaneous. A broader perspective ensued, which allowed me to experience a more whole and subtle understanding. As the letting go unfolded, I felt more intimate with sensory experience, more intuitive . . ."

Fehmi found that imagining space was one of the ways to force the brain to stop grasping and move into open focus. The state is experienced as "a vast three-dimensional space, nothingness, absence, silence, and timelessness. The scope of our attention is not only expanded, but is experienced with greater immersion. Thus, the ground of our experience is reified, realized as a more pronounced sense of presence, a centered and unified awareness, an identity with a vast quality-less awareness in which all objects of sensation float, as myself." This sounds surprisingly similar to the meditators' reports when they quieted the orientation area in their brains. You can get a taste of open focus now, if you want. As you read, become aware of the space in between the letters on the page while you are attending to the words and the meanings of the words. Can you also be aware of the space between you and the paper? At the same time, is it also possible to be aware of sounds around you? Let all of that stay with you as you attend to the words and to the meanings of the words you read.

Fehmi believes that the way we pay attention is important. If someone is always in narrow objective focus, he will start to experience stress, regardless of the content of his attention. Fehmi was chronically in narrow focus; that is why he experienced such a profound breakthrough. He finally gave up and went into the open focus state. Consideration of our society's chronic narrow focus may help us to explain both rampant drug use and fascination with meditation and ecstatic spiritual states. These methods help us alleviate the tension of remaining chronically narrow focused in our consensus trance.

The relief that comes with altering our attention and our consciousness is more than just feeling good. Fehmi's open focus, hypnotic trances, and other ecstatic states have been shown to bring about the remission of many stress-related symptoms, chronic pain, insomnia, even eye and skin disorders. People who have been the most narrow focused may experience the most profound results. With practice most people can experience lasting changes.

Though many of these changes are subjective and hard to measure, some studies are showing how our attention may physically change the brain. Susan Greenfield has shown how the hippocampi of London taxi drivers were enlarged in proportion with the length of their employment—possibly related to their remembering abilities. She also noted a similar study where just practicing five-finger piano exercises for five days enhanced the area of the brain relating to the digits. More remarkable is that just imagining the movements creates a comparable change in the brain, a measurable physical change.

Since we can demonstrate that imagination does change the structure in the brain, then it becomes more believable that an altered state can generate other paranormal phenomena. The ability to control pain and resist burning that firewalkers and hypnotized people display may be a natural, though seldom used, potential of the mind-body connection. Parapsychologists Russell Targ and Jane Katra say that the interconnectedness demonstrated in quantum physics is the explanation for psychic abilities like remote viewing and distant healing. Our ability to control our brains and minds puts us in touch with the experience and phenomenon of no separation. This is essentially the same thing the mystics have been reporting for thousands of years: that the separation between mind and body, between ourselves and others, even the phenomena of space and time, are illusions. Fehmi's attention training along with meditation and other consciousness-altering practices may be more psychologically and physiologically powerful than we've believed. Targ and Katra say, "The choice of where we put our attention is ultimately our most powerful freedom. Our choice of attitude and focus affects not only our own perceptions and experiences, but also the experiences and behaviors of others."

If you've been using your attention to alter your awareness while reading this article, you may have a sense of how easy it is to shift your consciousness. Your experience may seem totally unlike the possession trance of a firewalker or a shaman, but it is related. It's only a matter of degree. If you can fully appreciate the value of these milder controlled states, you may be more open to the value of states more alien to you, more altered. Those more extreme states have been used for millennia by indigenous shamans and healers to fulfill valid personal and social needs. Even scientists like Edison and Einstein used their ability to slip into natural trance states for creative breakthrough. Einstein even said some of his formulae were not derived from research or calculation, but from "psychical entities as more or less clear images."

Many meditators, hypnotic patients, and open focus practitioners who use these ASCs report they feel more in control of their lives. Their direct experiences from these states give them a flexibility that loosens the hold consensus trance has on their minds. A century ago William James said, "[T]he mystical feeling of enlargement, union, and emancipation has no specific intellectual content whatsoever of its own. . . . We have no right, therefore, to invoke its prestige as distinctly in favor of any special belief." The work of these brain and mind researchers is helping us to understand without needing to believe; both the physiological and the psychological knowledge we possess has value. It is freeing to realize that we need not be believers or skeptics but can explore our states of consciousness with a more flexible and clearer mind. We may enjoy and even be surprised at what we find.

37 Searching for the Unifying Field

Author Lynne McTaggart Is Tracking the World's Most Exotic Research toward Discovery of a Secret Force

Cynthia Logan

Investigative journalist Lynne McTaggart may look like a pixie (she could easily be cast as Peter Pan, Tinker Bell, or Puck in a *Midsummer Night's Dream*), but don't let the impish smile fool you. Her large, warm eyes probe deeply, seeing what others miss. Her keen interest in a variety of subjects, coupled with inherent intelligence, determination, and the ability to convey complex information with clarity, comprises a professional profile that has served her (and her readers) well. Plus, she's always up for a good role. In the 1970s, she posed as an unwed mother to get the story for her first book, *The Baby Brokers: The Marketing of White Babies in America,* an exposé of gray market adoption in the United States. "I spent several years uncovering international baby rings," she says of the "scoop" that would have intimidated less intrepid reporters. Her next book, *What Doctors Don't Tell You,* exposed practices the medical establishment would rather you not know, and her latest offering, *The Field: The Quest for the Secret Force of the Universe,* reveals revolutionary scientific discoveries that will change your life forever.

Interested in writing from the age of seven, McTaggart remembers being impressed by Bob Woodward and Carl Bernstein, authors of *All the President's Men,* recognized for exposing the Watergate cover-up. "As a teenager, I watched them take down a president—it showed the power and obligation of the press to investigate thoroughly, to be the guardian of human rights." Born in Ridgewood, New Jersey, McTaggart has always loved a good story, and was also influenced by authors Tom Wolfe and Joan Didion, whom she says combine narrative with nonfiction, exhaustive research, and "saturation reporting." After attending Northwestern University's Medical School of Journalism, she transferred to Bennington College, where she studied

Fig. 37.1. Author and researcher Lynne McTaggart.

literature and took advantage of the nine-week "work terms" offered as credits. "I had a job as an editorial assistant on *Playboy* magazine in the days when they carried good nonfiction. This was in the early seventies, when there were a lot of big, breaking stories. The next year I worked on the *Atlantic Monthly*—two very different experiences!" Postgraduation, McTaggart was managing editor of the *Chicago Tribune/New York News* syndicate and contributed to the *Saturday Review.* "The stories about health fired me up," she says. "I became a holistic detective."

After she moved to England, in the mid-eighties, McTaggart became mysteriously ill and put her detective skills to work. "I went from ordinary doctors to the outer rim of alternative medicine and no one seemed to get me better," she recalls. "I realized I'd have to research it myself and find the right person to help me." She was thrilled to eventually find a pioneer in nutritional medicine. "We worked in partnership to heal my condition" (*Candida albicans* overgrowth, now well understood), she says. With her husband and business partner, Bryan Hubbard, McTaggart began publishing a newsletter called *What Doctors Don't Tell You* (the book came out in 1999), still going today and extremely popular. "It's very research based," she says. "We scour the medical literature. It's filled with unbelievable stories about how their tools don't work." They also publish *PROOF!* which examines the evidence for alternative medicine. "We subject alternative medicine to the same scrutiny with which we look at conventional medicine," she says. This includes sending "branded" products to independent labs to determine whether they do what they claim to do. "We seem to be a little encyclopedia on what does and doesn't work," says McTaggart of their newsletters' growing subscriptions.

In the course of this work, McTaggart continuously "bumped up against solid scientific studies showing good evidence for homeopathy, acupuncture, and spiritual healing." Though a believer in alternative/complementary medicine, she wondered where "subtle energy" comes from, and whether there were such things as human energy fields. She reasoned that "if something like homeopathy or distance healing worked, it contradicted everything we believe about our reality. The debunker in me was left dissatisfied." Thus, she turned her formidable talents to the field of science, secured "a very sizable advance" from publishing giant HarperCollins ("it helped that we sold it in three countries before it was published"), and set about meeting and interviewing fifty "frontier" scientists. "I began a personal quest to find out if any frontier work in science could offer an explanation. I traveled around the globe, meeting with physicists and top scientists in the United States, Russia, Germany, France, England, and South and Central America. Their theories and experiments added up to a new science, a radically new view of the world."

She attended meetings and studied journals, digesting their concepts (reading and deciphering hundreds of scientific books and papers!) and learning their parlance. "Most scientists don't like to stray beyond their experimental data and don't like to synthesize," she notes. "They take baby steps and it's often difficult for them to see the big picture. They communicate in equations, the language of physics. I wanted to tell their stories and sneak the science in." *The Field: The Quest for the Secret Force of the Universe* represents one of the first attempts to synthesize disparate research into a cohesive whole. In it, says popular author Dr. Wayne W. Dyer, "McTaggart presents the hard evidence for what spiritual masters have been telling us for centuries."

The "field" referred to is the zero-point field, a subatomic realm of quantum energy existing in what has heretofore been considered a vacuum, empty space. (Actually, scientists have known about the "ZPF" factor, but have subtracted out this extra quantum energy in calculations because they felt it wasn't important.) This is the "unified field" spoken of by Deepak Chopra and is, says McTaggart, "a bit like 'the Force' in *Star Wars*." Consisting of the micro-movements of all the particles in the universe, it's a vast, inexhaustible, supercharged energy source hanging out in the background of the empty space around us. "To give you some idea of the magnitude of that power," offers McTaggart (whose practical, creative analogies provide solid stepping-stones throughout the book), "the energy in a single cubic yard of 'empty' space is enough to boil all the oceans of the world." With a fossil fuel energy crisis looming, scientists at top-ranking universities like Princeton and Stanford, and at many prestigious institutions in Europe, have realized that the zero-point field has enormous implications.

Astrophysicists have called the ZPF a "cosmic free lunch." If successful in harnessing this source of limitless energy, they may be able to create antigravity WARP drives and cars that run without fossil fuel. We may travel beyond our own solar system within the foreseeable future—both NASA and British Aerospace are researching the possibility of recycling the energy in empty space. Besides, McTaggart points out, the little bad guy in *The Incredibles* mentions zero-point energy! Perhaps more important, though, she thinks the existence of the zero-point field implies that all matter in the universe is interconnected by quantum waves, which are spread out through time and space and can carry on to infinity, tying any one part of the universe to every other part. Like an invisible web, the field connects everything in the universe. "The idea of the field might just offer a scientific explanation for the spiritual belief that there is such as thing as a life force."

She finds this probability comforting: "It's as though, on the tiniest level of reality, a memory of the universe for all time is contained in empty space that each of us is always in touch with," a view far more nurturing than the reductionist, separatist paradigm we've inherited from modern established science. While she became relatively comfortable with scientific concepts writing *The Field,* McTaggart nevertheless had some hurdles to jump. "The most difficult idea I had to get my head around was the notion that this giant energy field operates outside of space and time, and that humans can have effects on the world around them outside of space and time. It appears that since subatomic particles can interact across all space and time, so does the larger matter they compose. The idea that there is absolute time and space has to be replaced with a truer picture of a universe that exists in one vast 'here,' where 'here' represents all time and all space at a single instant."

At our most elemental, writes McTaggart, "we are not a chemical reaction, but an energetic charge. All living things are a coalescence of energy in a field of energy . . . this pulsating energy field is the central engine of our being and our consciousness, the alpha and omega of our existence." The field is, in other words, our brain, heart, and memory. Many scientists now postulate that the human brain is more like a radio than a computer. McTaggart's own research indicates that "a load of environmental issues gum up the radio, clogging the receiving mechanism." She notes the connection between environmental pollution and memory loss. "Many pesticides have negative effects on the brain . . . and there is a decent connection between mercury in amalgam fillings and Alzheimer's."

Of all the physical and metaphysical implications of these multiple discoveries, none are more exciting to McTaggart than those that point to changes in health care. This is, after all, her passion. "Within 50 years, the idea of using drugs or surgery to cure anyone will seem barbaric," she says. Instead, she thinks we'll be manipulating people's quantum energy, as a number of researchers using cutting-edge techniques are attempting to do already. She cites a group of French doctors using molecular frequencies to identify specific bacteria and pathogens, and another in Germany measuring quantum emissions to determine the quality of food.

The McTaggart/Hubbard household is located near Wimbledon, where McTaggart, self-described as "relatively sporty," likes to ride her bike and swim. The couple has two daughters, Caitlin and Anya, and the family is closely knit. "We don't work on weekends," she says, "we do a lot of things together." Their whole, unprocessed, organic-foods diet is sprinkled with Chinese take-out and (for the parents) a couple of bottles of wine a week—"we believe in the fun factor," she says with a laugh. "I'm fairly strict about not having wheat in the house. We take loads of supplements since the evidence is pretty clear that the nutrition in food isn't as good as it used to be."

Recently, she, Hubbard, and their company, WDDTY, spearheaded the Health Freedom Movement, a nonprofit organization dedicated to fighting European and international laws threatening freedom of choice in natural medicine. Like the Royal Family, they turn to homeopathy and acupuncture for preventive health care. "The alternative therapies are alive and thriving in England," reports McTaggart, whose dentist not only avoids toxic mercury fillings but also understands the Chinese meridian system and how teeth relate to that network. "You're allowed to be eccentric in this country—I have an amazing network of alternative practitioners." According to McTaggart, more people visit alternative practitioners in England than go to conventional doctors, and a full third of British citizens use nontoxic cleansers in their homes. "The British medical system is bankrupt because everyone believes they should get health care for free," she remarks. "It's great to have health care for all, but there are long waiting lines."

In contrast, McTaggart is encouraged by waiting lines that have formed for the independent film *What the Bleep Do We Know!?* wherever it's been shown—to her mind, a further indication of a positive shift in the general consciousness. The movie brings viewers face-to-face with quantum physics' implications and possibilities, and features interviews with leading experts and scientists—just the sort of thing she'd like to see more of.

38 The Mind-Matter Connection

Did the Long-Sought Smoking Gun of Scientific
Proof Finally Appear on September 11, 2001?

John Kettler

It has long been held, by metaphysicists, occultists, and many theologians alike, that there is no such thing as a mere thought: that a thought is a thing and exerts real power, a view some physicists are beginning to adopt as well. Thus, in esoteric and theological circles we find endless warnings against unguarded thoughts, wrongly directed thoughts, and deliberately misused thoughts. We see one discipline and practice after another designed to first center the individual, clear the mind, and enable proper focus on the desired thought or even contemplation of the void, a state of no thought in a completely stilled and empty mind.

At the opposite end of the spectrum we find the highly agitated, excited mass mind, a mind anything but centered and at peace. It is this mind, whether naturally occurring or artificially created, that is the special province of politicians, propagandists, ad agencies, public relations firms, con artists, and mental health professionals of every type and ethical stripe, not to mention an array of covert operatives reporting to all sorts of governments, groups, and individuals. Interestingly, it has also emerged as an unexpectedly fruitful area for psi research on mass populations.

How? To answer that question, we must first delve a bit into how psi research was conducted long ago and the path it has followed since then.

RANDOM NUMBERS AND PSI

The great quest of psi research—repeatable, documented proof of the phenomena—hinges upon first constructing and conducting an experiment so airtight in every single aspect as to exclude all other sources of potential error, accidental or otherwise. To this end, there has been an ongoing effort to develop a truly random generator of events, events that can then be tested via rigorous statistical means for evidence of psi phenomena in the form of someone's attempting to predict or even influence them.

The early work, under such pioneers as J. B. Rhine at Duke University in the 1920s, was based upon the calling of manual coin tosses and card draws, processes later automated and ultimately computerized. Such work also led to the development of Zener cards, the ones with the familiar stars, crosses, and the like, with the specific goal of giving the subjects trying to telepathically "send" the cards a strong, distinct mental image to convey to the person acting as the telepathic "receiving unit."

The advent of computers changed everything, though, by greatly increasing the speed and numbers of tests that could be conducted and the rate at which the resultant piles of data could be analyzed using statistical methods. Computers were themselves used as random number generators (RNGs), but it was soon found that the numeric "seed" used to key the RNG sequence was itself affecting the randomness of the numbers being generated. This led to a search for truly random phenomena, which could be harnessed for the painstaking work involved, one of which turned out to be radioactive decay.

Years ago, public TV's *Nova* aired a program on psi in which an RNG operated by that decay was used to determine the direction a light in a ring of unlit lights would go. The decay rate of the material was known, but the precise instant of a given decay event was unknown, and it was this unknown timing that was used to randomly "drive" the light around the ring either clockwise or counterclockwise. The task of the test subject was to make the light obey his will by causing it to move in the direction the experimenting scientist commanded. Amazingly, the subject proceeded to do just that, marching the light at will in the specified direction for a time before tiring.

Radioactive materials have inherent safety, legal, and administrative drawbacks, so there has recently been a migration to specialized, electro-magnetically shielded RNGs, which utilize several types of exotic events (resistor noise and quantum tunneling) as the trigger for generating their random numbers. There are about fifty (a few less or more depending upon computer maintenance issues) such devices in use worldwide for scientific research under the aegis of the Internet-based Global Consciousness Project (GCP), and their output is under the most minute scrutiny at all times. Theoretically, they are immune to human and other forms of intervention when operating properly and not having been tampered with, but several traumatic global incidents may have demonstrated en masse what generations of individual researchers couldn't do in their labs—the influence of focused consciousness upon the real world.

Fig. 38.1. Dean Radin, senior scientist at the Institute of Noetic Sciences and explorer of psi phenomena.

In that influence, properly understood, lies the very essence of metaphysics, ritual magick (to distinguish it from stage "magic"), and prayer—using nothing but thought energy to affect this earthly reality. Enter researcher Dean Radin of former astronaut Edgar Mitchell's Institute of Noetic Sciences, Petaluma, California, and his amazing paper "Exploring Relationships Between Physical Events and Mass Human Attention: Asking for Whom the Bell Tolls" in the *Journal of Scientific Exploration,** which may well provide the long-elusive smoking gun for psi phenomena.

WHEN THE RNGS WENT TILT!

Dean Radin's paper was but the latest in an ongoing series of investigations into the connection between mind and matter, with the earlier work being devoted to investigating the effects before, during, and after of "highly focused or coherent group events" on the output of "electronic noise based, truly random number generators (RNGs)."

His conclusions may startle: "Results of these studies suggest in general that mind and matter are entangled in some fundamental way, and in particular that focused mental attention in groups is associated with negentropic [antirandom] fluctuations in streams of truly random data." In other words, focused mental attention of groups apparently is affecting the very randomness itself of the output from the RNGs, a theoretically impossible occurrence given their design and implementation.

No better test of this premise can be imagined than what befell the United States of America and the rest of the world (some seventy countries lost people when the World Trade Center was hit twice and quickly col-

---

*Journal of Scientific Exploration 16, no. 4 (Winter 2003): 533–47.

lapsed) on September 11, 2001, and that was but part of the horror on a day that saw thousands of traumatic deaths and scenes of incredible devastation broadcast worldwide almost instantaneously. Dean Radin's hypothesis was that such a catastrophe would in fact affect the RNGs' data streams, and was he ever right. His painstaking research and analysis shows that September 11, 2001, marked the single biggest negentropic episode to occur in the entire year of 2001. Further, it occurred not in a few isolated RNGs, but was recorded by the entire network, which is global, after controlling for a handful of RNGs that were malfunctioning. Only those RNGs operating within their proper parameters were analyzed. In order to be accepted, each device underwent a series of grueling tests, including a calibration test consisting of "one million 200-bit trials."

It's one thing to assert or claim that September 11, 2001, was the single largest negentropic fluctuation measured by the Global Consciousness Project for the entire year of 2001. It is quite another, with much of the proof hinging upon a gamut of meticulous statistical tests, to say that the event *caused* the observed result, for there may well be other factors at work. Did the researcher, say, happen to choose an event window duration that caused

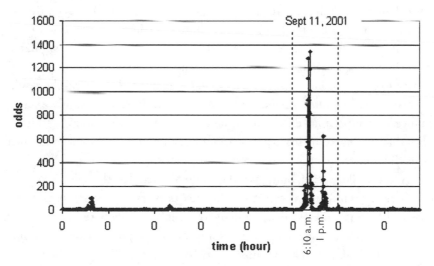

Fig. 38.2. This image illustrates the two-tailed probabilities associated with the smoothed Z score as odds ratios. There is an extraordinary spike near the time of the attacks, driven by large deviations that precede the first plane crashing into the World Trade Center tower; its weighted center is at 06:10, corresponding to the peak in the Z scores. The second spike occurs roughly seven hours later, with the weighted center at about 1:00 p.m. The "0's" in the x-axis show the start of each day. (Image courtesy of the Society for Scientific Exploration)

data to display anomalously under that set of conditions and no other? Dean Radin addressed this matter by testing all kinds of window lengths. Was it an artifact of the sampling procedure? He reran the tests using a range of sampling methods, but the results didn't change. The same held true when he tried to find out whether unusual environmental conditions, diurnal conditions (day–night, with their implication for electronic interference), or even cell phone usage might be skewing his data. Nothing was found to account for the obviously marked departure from the statistical norm for the behavior of the GCP network's RNGs.

To make sure that he wasn't fooling himself by using knowledge obtained after the fact, he then went through multiple lists, day by day, of events reported for the year 2001 by several news services, noting what event got how much coverage. Armed with this additional information, he then went back and applied the same statistical approach to other mass attention events, such as Princess Diana's funeral, and found that the predicted outcome matched the observed outcome. In other words, events other than 9/11 that drew strong mass attention also generated negentropic fluctuations in the RNG network, though to a lesser degree.

Contrary to what some may think, scientific investigations don't always immediately yield answers to the questions they raise. In fact, even though we know how to manufacture and distribute electricity, we still don't really understand it after centuries of study. Dean Radin's situation is similar in that he believes he's found an anomaly that appears to conform to the mind-matter hypothesis. He now needs to go back to his fellow researchers with his results and conduct further tests, in order to more thoroughly exclude chance or the peculiar effects of some obscure analytical choice. While he does that, we have the luxury of leaping ahead and pondering the implications of what seems to be a staggering discovery.

At a very deep level, humankind has always acted as though there was a link between mind and matter, a fact borne out by rituals dating back to the dawn of history and candlelit vigils held tonight somewhere in the world. We storm out into the streets in riots and at other times picket peacefully, standing up publicly for our most cherished beliefs. We come together in common cause in natural places and man-made structures as we wrestle with our place in the scheme of things, to worship, obtain solace, and ask for what we need, to focus our attention on individual and group goals. We even have Internet-based prayer circles that girdle the globe.

Perhaps now is the time to harness the power of the collective mind for the collective good.

39 **Dr. Quantum's Big Ideas**

Fred Alan Wolf Searches for Concrete
Answers to Ephemeral Questions

Cynthia Logan

oogle the name "Fred Alan Wolf" and you get the semi-astronomical figure of 2,410,000 responses in 0.24 second—an appropriate representation for the theoretical physicist who appeared in the runaway indie film *What the Bleep Do We Know!?* and who calls himself "Dr. Quantum." Also a writer and lecturer, Wolf earned his Ph.D. at UCLA in 1963 and subsequently a reputation for simplifying science by putting complex concepts into layman's terms. His book *Taking the Quantum Leap* won the National Book Award (1982) and is still selling nearly as well as it did when first published ("my book, unfortunately, is probably one of the best ones out there. I hate to say it—it's especially good for people open to the mystical or consciousness part of it." It's been listed by the American Library Association as one of the top books ever—*ever*—written on science!). Wolf is also the author of *Parallel Universes, The Dreaming Universe, The Eagle's Quest, The Spiritual Universe, Mind into Matter, Matter into Feeling,* and *The Yoga of Time Travel: How the Mind Can Defeat Time.* His latest book is *Dr. Quantum's Little Book of Big Ideas.*

Having taught at San Diego State University, the universities of Paris and London, the Hebrew University of Jerusalem, Birkbeck College, and Hahn-Meitner Institute for Nuclear Physics in Berlin, Wolf was well known in academia for his contributions through technical papers. He is now in demand as a lecturer, keynote speaker, and consultant to both industry and the media. In addition to his memorable appearance in *What the Bleep!?* (remember the very animated, slightly balding guy with a neatly trimmed grey beard and glasses?), Wolf has appeared as the resident physicist on The Discovery Channel's *The Know Zone* and on television and radio talk shows across the United States and abroad.

Calling himself "an introvert playing the part of an extrovert," Wolf has performed both as a stage and a "close-up" magician, as well as playing the harmonica in front of an audience. "Part of me is very much an

Fig. 39.1. Physicist, writer, and lecturer Fred Alan Wolf.

entertainer," he admits. This comes across even on audio media, where his energetic message is delivered with punches of inflection that, while at times annoying, convey his points both dramatically and emphatically. The notion (this is a word he uses frequently) of creating a memorable figure to help people grasp the complexity of theoretical physics came in the early 1980s, when Wolf and his cousin came up with "Captain Quantum," who attended conferences attired in a cape. *Future* magazine created a cartoon character of Wolf as the captain. Then, when *The Bleep* came out, the producer wanted to use it, too. The moniker, however, had been trademarked for a board game, "so it was changed to 'Dr. Quantum,'" says Wolf, who co-owns the new trademark with the film's producers. The expanded *Bleep* movie, *Down the Rabbit Hole,* has Dr. Quantum traveling around "doing cartoon kinds of things." Version two, according to Wolf, is forty-five minutes longer than its prequel. "It's the same storyline, and no acting parts were added," he reveals. "Some speakers' information has been updated and other speakers have been added."

The film's central tenet is that we create our own reality through consciousness and quantum mechanics, a theme Wolf endorses with a caveat. "The 'new-age' thinking that 'whatever happens to me I've attracted to myself' is misleading and somewhat unfortunate, because there's so much more than that," he explains. "It's more than the 'motivational speaker mantra' (you've got to get up and do your thing) that's unimportant relative to the deeper message. I try to teach the depth of quantum physics." His latest thoughts, encapsulated on CD in *Dr. Quantum Presents: A User's Guide to Your Universe,* address the basics of quantum mechanics, the nature and role of consciousness, the possibility of parallel universes, the imaginal realm, time traveling through the universe, and sex, magic, and the shamanic world, among other topics.

Wolf feels the movie's success stems from a growing shift in the mood of the country toward a middle ground. "Neither the extreme left or right would find it appealing," he notes, adding that "being from the Midwest may have contributed to my own tendency to see things from the middle." Growing up in Chicago, his fascination with physics began one afternoon when, around the age of ten, he witnessed the world's first atomic explosion at a local matinee. Though his major interest in sports lasted through high school, he went on to study mathematics and physics at the University of Illinois, then attended UCLA, where, as mentioned, he took his Ph.D.

Having participated in master classes with Nobel laureate Richard Feynman, and conversed with one of the fathers of modern physics, Werner Heisenberg (winner of the Nobel Prize for his formulation of the Uncertainty Principle), Wolf is currently researching a model of consciousness that investigates the nature of observation. Are some people better observers than others? This relatively new "weak measurement theory" still fits within the framework of quantum physics, but indicates we can model what an observation is and whether it is strong or weak. According to Wolf, weak observations turn out to give us results that are affected by what will be measured in the future; in other words, future measurements seem to have an effect on present observations.

Working with mathematical formulations on questions like this from his home in San Francisco, Wolf ponders the relationship among human consciousness, psychology, physiology, the mystical, and the spiritual. He points to dialogues held in ancient times, explaining that there was no separation then among philosophy, religion, and spirituality. "The Greeks talked about earth, air, fire, and water," he mentions. "They also talked about a quintessence ["fifth essence"], which they called *physis*, which was the spiritual aspect of it all, from which the word *physics* even comes. So it seems to me that in our present state of consciousness, that kind of bridge could be made, and be fairly firm."

Constructing and walking across that bridge is what he's about, and he likes to keep it simple and concrete. He's earthy ("I don't value a 24-carat diamond—I can't eat it, I can't even wipe my butt with it"), yet believes the physical world has a spiritual basis ("things change as a result of observing; that leads to the question of what is meant by observing . . . what is mind? Once you get into mind, you can't stay in the physical realm, because mind is not brain"). The only answers he's found that deal with what he's discovered in quantum physics exist in the vast wealth of literature of the ancient mystics. "I don't find anything in the literature of modern mystics; they're overly complex and their ideas stray from the basic ideas I find in quantum physics. The new guys are off spinning their own stories and don't relate back to science." On the other hand, he notes the host of new literature written by "highly intelligent philosophical/scientific thinkers circulating the idea that maybe the answer is that we really don't know." His game, he says, "is that we have a science that has introduced a whole new notion of what our world is, and there's no way to avoid the question of observation affecting reality. It's non-mechanical . . . something magical, with an awe and wonder about it. To neglect that is to neglect part of the awe and wonder of being alive."

While he's living life to the fullest (he enjoys his senses in both waking and lucid dreaming states and has taken the famous "fire walk"), he finds that all spiritual traditions have a basic underlying notion (the N-word again) that this reality is illusory. The path is to wake up from the dream we're walking in. And while "some people think we'll be out in the cosmic boondocks in Nirvana," he doesn't see it that way. But rather than expounding on his beliefs (not knowing is, after all, the closest thing he espouses), he states succinctly: "Religion is a kind of a vessel that hopes to hold a spiritual elixir, but mostly it's just an empty vial."

In his view, most people don't receive the enlivening spiritual experience that religion is supposed to give them. His own definition of a spiritual experience is akin to the "aha!" phenomenon—that sudden lightning bolt of enlightenment. "I've had many such awakenings," he relates. "A lot of them have occurred during my travels to other parts of the world. Once, at a Buddhist temple in India, I had a spiritual awakening, believe it or not, when a fly landed on my foot. While the Buddhists were chanting, a fly suddenly landed on my foot. I felt as if my consciousness and the fly's had become one. When I looked down to see where the Buddhists were chanting, I saw an infinity of Buddhist monks going back all the way to the beginning of time. It was like looking through an infinity mirror and it all happened in a flash, and was very moving to me."

Wolf had a close relationship with the late physicist David Bohm: "My office was next to Bohm's when I worked at Birkbeck College in London. He'd come in and say he wanted to talk and, basically, you had to listen. I saw him change after he had a heart attack. Six months or so before he died, he seemed to have a mystical quality; he had practiced surrendering, something we all have to do before we die." According to Wolf, many scientists experience a spiritual awakening when somebody close to them dies.

"Suddenly, they realize what life is about, and they begin to see the illusion of living forever—they're open to the possibility that your head may be full of demons when you think all that's out there is a godless universe of law and chaos."

Death is something Wolf himself has thought about since both his son and his mother died within six years of one another. He thinks it may be a return to what he calls "the Big Elephant." "It's a funny thing—spirituality is like an elephant in the room, a huge thing that nobody can see. The elephant is your spiritual essence, your essential self. That's different from your ego self, or the person you identify with in body/mind consciousness. This other form of consciousness may actually be running the show, and we have no idea who or what it is. The evidence seems to be pointing to the conclusion that there is only one true observer in this whole universe. And what death seems to be is a return to that one observer—whether you want to call that God, the soul of the universe, or just the Big Kahuna, I don't care. But that's what seems to happen."

In spite of all the death and destruction in the world, Wolf thinks it's getting better. In his view, people are overly pessimistic. "Things have gotten better, though the fears have gotten worse. I think a lot of the improvement has come through our sensible management of money and trade; I think we're moving in the right direction. Trade barriers have to come down—I don't know if that's Republican or Democratic—but we're moving into an international society and I think that's good."

Having traveled extensively, he notes radical, positive changes in both India and Mexico. His book *The Spiritual Universe* has been translated into two Chinese dialects and has been well received in that huge, emerging market. Not everyone, though, appreciates his take on the world.

Michael Shermer, a columnist for *Scientific American* and a professional critic, comes to mind: "I like him personally, though he doesn't agree with anything I say." In an essay titled "Quantum Quackery," Shermer went after *What the Bleep Do We Know!?* with humor and vengeance. "The film's avatars are New Age scientists whose jargon-laden sound bites amount to little more than what California Institute of Technology physicist and Nobel laureate Murray Gell-Mann once described as 'quantum flapdoodle,'" he writes.

Bring it on, invites Wolf: "I find skeptics are interesting to talk with." While he and University of Oregon quantum physicist Amit Goswami agree on most points, even they have their differences, and Wolf notes that most scientists have "language problems" (Dr. Candace Pert, credited with discovering the brain's opiate receptor and author of *Molecules of Emotion,* has quipped that scientists would rather use each others' toothbrushes than each others' terminology). "We tend to get caught in loving ourselves and our own ideas rather than someone else's," says Wolf. "We all do it; you can't really be a scientist without doing this." Wolf not only understands his critics, but embraces them as well. "I think a lot of the things I say should be looked at skeptically. I don't claim I have absolute truth here, I'm just putting out ideas, trying to be as clear as I can." And whether or not you agree with his "matrix of possibility," Dr. Quantum offers plenty of food for thought and one thing's for sure: it's neither bland nor boring.

PART TEN

THE FUTURE

40 Further Explorations of the Crack in the Cosmic Egg

Joseph Chilton Pearce's New Book
Probes the Biology of Transcendence

Cynthia Logan

Think those pesky telemarketers are annoying? While they (not to mention those computerized calls!) can irk even the most polite among us, they're tame compared to what one family endured over fifty years ago. Awakened at midnight by an incessant knocking at the front door, the lady of the house finally opened it, finding now famous author Joseph Chilton Pearce (*The Crack in the Cosmic Egg, Magical Child, Evolution's End*) standing there. Pearce was peddling silverware and, buoyed by a state he calls "unconflicted behavior," he felt a power he admits was then misused.

As an irate mother, sleepy daughter, and bewildered father stood around their dining room table gazing at a magnificent spread of silver, Pearce found his adrenaline pumped, sure that a sale was imminent. "At each new outburst from the wife to her husband to 'throw out this little mouse!' sheer exhilaration and excitement welled up in me and I began to laugh until tears streamed down my face," he recalls. "The more I laughed, the angrier the mother became and the more bewildered the other two looked; the more surely they all lost control, the more vulnerable they became." As he walked out with a large order and a down payment, Pearce also walked arm in arm with the parents, who implored him to come back and visit them! "I realized that the average person in his or her conflicted state of uncertainty, doubt, and fear (which was my ordinary state as well) was not only powerless in the face of unconflicted behavior, but was also seriously attracted to this state."

Pearce had come to know that "the structure of reality is negotiable" at twenty-two, when he had three "blackout" experiences during which he found himself in a location other than that of his body. Later, while attending university classes during the day and working an eight-hour graveyard shift six nights a week, he found he could access the "unconflicted" state at will. Essentially, his wide-awake body ran the check-proofing machine he was assigned to operate while his consciousness was elsewhere, affording him desperately needed sleep. This arrangement was stellar: of the many workers required to process thousands of checks each night, only Pearce produced work completely free of error!

Such experiences (among others) fueled his seventies bestseller, *The Crack in the Cosmic Egg,* which he began writing in 1958 and sold in 1970. The "crack" refers to what Carlos Castaneda calls "a cubic centimeter of chance" and what Deepak Chopra refers to as "the gap," a brief (we're talking a nanosecond) window of time when we can slip into a state of mind that affords us a transcendent perspective. It seems paradoxical, but Pearce says that by accepting death completely we have access to a vast realm of life that great beings throughout time have described. Further, in his latest work, *The Biology of Transcendence,* he claims that we are literally made to transcend our current evolutionary capacities and limitations, and explains the dynamic interaction of what he terms the "head brain" (intellect) and the "heart brain" (intelligence): "Intelligence is connected with the deepest intuitive roots of life, the matrix of our being.

"It is the dark, mysterious interior of life. And intelligence is essentially feminine in its nature, if I may use that kind of polarity. It is subjective and interior. On the other hand, intellect is objective. It is exterior and outwardly driven. Intellect is brain-centered, analytical, logical, linear. It is constantly questioning. It takes things apart and loves to put them together in new ways. It is inventive. This brain-centered, objective intellect is a highly specific form of intelligence that is evolution's latest achievement." Yet, he says, it is meant to serve in tandem with the heart and can be thwarted if lower brain structures require more attention than nature intended.

Current biological understanding of brain structure includes four neural centers: the hind, or "reptilian"; the middle, or "limbic" (so named for its limblike attachment to the lower brain); the forebrain, or neocortex; and the prefrontal lobes, which Pearce claims carry our capacity for transcendence. The emerging science of neurocardiology postulates a fifth center in the heart.

Long interested in the heart as "infinitely more than a pump," in 1995 Pearce came across HeartMath Institute in Boulder Creek, California. "They had been gathering research information from around the globe, including a huge, thick volume of medical studies from Oxford University entitled *Neurocardiology,*" he says. "Discoveries in this field are, believe me, far more awesome than the discovery of non-locality in quantum mechanics!"

Pearce outlines three major tenets of the new discipline: 1) 60–65 percent of all heart cells are neural cells, identical to those in the brain; 2) the heart is the major endocrine glandular structure of the body, producing hormones that profoundly affect the operations of body, brain, and mind; 3) the heart produces two and a half watts of electrical pulsation with each beat, creating an electromagnetic field identical to that of the earth. His *Biology* includes many "HeartMath charts" and explains in detail the procedure known as "Freeze-Frame," a technique designed to access the intelligence of the heart when we're under stress.

Pearce feels that accessing this intelligence is vital in the choice we have either to achieve transcendence or to perpetuate the cultural violence that threatens to annihilate the planet. "Our violence toward ourselves and the planet is an issue that overshadows and makes a mockery of all our high aspirations," he says. "Recent studies reveal that the answer to cultural pathological violence is to be found in how the developing brain is encoded or programmed for peaceful or violent behaviors in the newborn/infant/child and by how it is birthed and raised." According to Pearce, the actual physiology of the brain shows a person's proclivities—the book contains a striking scanned image of a "normal" individual's brain and that of a violent person; the pictures are vastly different.

Born in Pineville, Kentucky, in 1926, Pearce was a passionate lover of the Episcopal Church throughout his childhood and "the hottest acolyte in the Southwest Virginia Diocese." Offered a scholarship to the best prep school in the South (to be followed by the university and then the seminary), Pearce's mother intervened. "She reminded me that ours was a family of newspaper people—that her brothers and father, like my father—had been editors and writers," he recalls. World War II intervened further, and Pearce spent his late adolescence in the Army Air Corps, openly weeping during what he calls a "Why-We-Fight" film that showed atrocities designed to incite future pilots and bombardiers toward "the mass murder required of all good airmen."

In spare moments he read Will and Ariel Durant's view of history. "In light of that wisdom and the horrors I viewed on screen, I dutifully became an atheist," he says. "Secretly, however, I held to a love of Jesus and a long-cherished romantic image of him, a kind of closet affair of the heart that had grown over the years. Of God I had my severe doubts; of Jesus as the greatest of humans and model for us all, I had none."

After the war, Pearce pursued higher education at an impressive array of institutions, attending Juilliard, Kings College, the L.A. Conservatory,

USC, William and Mary College, Indiana University, and Geneva Theological College, earning both his bachelor's and master's degrees in the humanities, which he taught at college level until 1963. Since then, he has written seven published books and has lectured at most major universities in the United States as well as many in England, Australia, New Zealand, India, Canada, Japan, Italy, Belgium, and Thailand.

Pearce became the single parent of four children after the sudden death of his wife at age thirty-five. After a profound paranormal experience that rendered him "nearly unable to function in

Fig. 40.1. Seminal thinker, author, and advocate of evolutionary child-rearing Joseph Chilton Pearce. Photo by Owen S. Peterson

normal life," a series of "miracles" enabled him to hold his university teaching post, attend to his four children, and continue work on his book. A subsequent marriage brought another daughter, now twenty-two. Today, living in the Blue Ridge Mountains of Virginia, Pearce enjoys twelve grandchildren and is grateful that many of them have been born at home, as were most of his children. Natural birthing, breast-feeding, and a Waldorf education (conceived by Rudolf Steiner, a visionary Pearce reveres), the only form of structured schooling he doesn't think should be completely razed, are crucial to spiritual development in his view, and something he has written about extensively. "Of my seven books, four center on child development, and have been adopted by various college courses," he says.

Indeed Pearce, a man of small stature, has been a giant champion of children and a leading voice in the human development movement—a voice in constant demand over the years. A faculty member at the Jung Institute in Switzerland, he spoke on "The New Paradigm of Human Development" at the seventh annual Transpersonal Psychology conference in India; Oxford University invited him to address the impact that current obstetrical practices are having on the development of intelligence; the Canadian government sponsored a workshop with Native Americans on the prevention of violence and substance abuse. Sony Corporation sponsored a seventeen-day lecture series on the future of education in Japan and Hawaii's crime-prevention commission asked him to discuss the current causes of crime and violence there; the state of Louisiana sponsored an address on the crisis facing the American family. Three different departments, at Harvard University, the University of California, and Stanford University, have each sponsored educational conferences featuring his work. The governor of California (pre-Schwarzenegger) requested Pearce address two special legislative planning sessions on the challenges facing children and families. And last year at Columbia University he spoke at a special conference on education in the twenty-first century.

Still, the man has time for play, something he deems of particular value and an absence of which he thinks is particularly disturbing as a cultural choice. He differentiates culture from society, explaining that culture is a "body of knowledge of learned survival strategies passed on to our young through teaching and modeling." As such, it is driven by our reptilian hindbrain, geared toward reflexive, primary survival instincts hardwired for defense. A society, on the other hand, includes what we often think of as "cultural": the arts; civilized, refined behavior; and so on. He reflects on what he considers a misuse of "socialization" in young children, particularly toddlers repeatedly told "NO!" when they are first exploring the world around them: "We experience deep conflict and internalize shame when hearing repeated negative commands from the person(s) we have come to trust implicitly."

As a result of such enculturation through negative programming, "all the news that's fit to print is generally negative news," notes Pearce. "Without a negative to induce our persistent focus, we won't pay attention, whether it be in the form of news, television, politics, economics, ecology, health, education or religion." He also believes that an absence of nurturing, affection, playful movement, and breast-feeding early in life results in a variety of brain abnormalities associated with depression, aggression, lack of impulse control, substance abuse, obesity, and violence.

Returning to the ("feminine") intelligence of the heart and reempowering women, especially with regard to reproduction, birth, and nurturing, is the "heart" of Pearce's hope for humanity's future. He calls it "the Resurrection of Eve," who, writes Glynda Lee Hoffman in *The Secret Dowry of Eve,* preceded Adam. "Any biologist would affirm that," quips Pearce, who also cites Plato's statement: "Give me a new mother and I'll give you a new world." Mothers fulfill what he terms "nature's model imperative."

"Development requires a model to trigger the brain to open a capacity.

If the model is not there, no development takes place. The character, nature, and quality of the model determines the character, nature, and quality of the developing intelligence," he says. "We must be who we want our children to become."

Though Pearce encountered and embraced Swami Muktananda, was involved with the SYDA Yoga Foundation for twelve years, and considers many enlightened individuals to have embodied transcendence, he returns to his childhood hero whenever he reaches for a description of what it is we're aiming for: "As model of a new evolutionary intelligence, Jesus met and continues to meet a grim fate at the hands of the survival culture that both spawns and is spawned by religion and myth. But the cross symbolizes both death and transcendence for us—our death to culture and our transcendence beyond it. If we lift the symbol of the cross from its mythical shroud of state-religion and biblical fairy tale, then the cross proves to be the 'crack' in our cultural cosmic egg, an opening to nature's new mind, wherein lies our true survival."

Though he doesn't have a website, you can encounter some of Pearce's ideas online via Michael Mendizza's website for *Touch the Future* (ttfuture.org—select "Joseph Chilton Pearce").

41 Challenging the Reality Consensus

A Popular New Movie Has People Thinking about the Unthinkable

Patrick Marsolek

The movie *What the Bleep Do We Know!?* has people talking about what constitutes reality and how our consciousness affects it. In this film, the strange realm of quantum physics serves as the launching point into an alternate perspective of the universe. The message of the film is simple: our consciousness does have a role in creating the reality we experience.

This message has attracted the interest of a wide range of people. Groups like the Unitarian Church, the Baha'i faith, and the Institute of Noetic Sciences see this movie as validating their beliefs. Conservative religions, mainstream scientific communities, and psychologists, on the other hand, believe the movie misrepresents science and is leading the public astray.

Let's look at how *Bleep*—and quantum physics in general—challenges the way we perceive reality and consensus beliefs. Later we'll give you a simple exercise you can do to challenge your perception of reality.

There are a number of key notions from quantum theory threatening the mainstream materialist view dominant in our culture. These are: the basic building blocks of reality are quanta, which are bundles of energy or information, not matter; reality is based on events, not things; quantum events are not causal—rather there is an innate indeterminacy and unpredictability to everything; events have complementarity and must be described as being both physical and energetic; and lastly, there is the strange quality of quantum participation—observation influences quantum events.

As demonstrated in the movie, the observer effect is most often used as "proof" by proponents of nonmaterialistic viewpoints—consciousness is important and not just a by-product of the brain. In quantum physics, consciousness and matter are connected. Scientific materialists—those whose beliefs in the primacy of matter are implicit and all-pervading—are very disturbed by this connection. The elevation of consciousness is largely why *Bleep* has generated such a strong response on both sides.

Quantum physics offers us a worldview that is radically different from what we've been taught to believe is real. Even individuals who are open to the new ideas in quantum physics have a difficult time grasping the implications of nonlocality, quantum participation, and the essential unpredictability of the universe. Considering these issues, as *Bleep* does, forces each of us to question the truth of the reality we collectively know. This line of inquiry pulls against everything we feel to be normal and can be very discomforting.

The psychologist and researcher Charles Tart coined the term "Consensus Reality Orientation" (CRO) to refer to our normal, day-to-day consciousness, which he describes as a kind of trance. Being aligned with this "consensus trance" allows each of us to live in our society, but it also constrains us to one perspective. The induction of the cultural trance is far more powerful and thorough than anything we ever do consciously with suggestion or hypnosis. Our current materialist view of reality has been spreading about the world for the past three hundred years. Individually, the consensus trance ingrained from birth continues through the entire course of our upbringing, culminating in higher "education." It's no wonder we are resistant to other worldviews. The scientists who've had the most "education" are often the most locked in to the CRO.

There are, however, always individuals promoting ideas that challenge

Fig. 41.1. Brain neural-nets with pictures of Amanda, as played by Marlee Matlin, from the movie What the Bleep Do We Know!?

Fig. 41.2. Author and teacher Charles Tart is internationally known for his research with altered states, transpersonal psychology, and parapsychology.

the consensus views, with consequences. Twenty-four hundred years ago, Socrates and his ideas were considered so dangerous that he was condemned to death. His student Plato asked us to imagine a cave where people were chained to a world of shadows. His parable is clearly a commentary on his society's CRO and highlights the difficulties of changing perspectives. Today there are still tremendous pressures to conform to the cultural CRO. Not so long ago, a scientist who dared venture into the realm of nonmaterial views was quickly blacklisted and discredited. Labeling a peer as crazy was just as effective as a sentence behind bars.

The internal difficulties of shifting out of the CRO can be just as strong. People experiencing near death, a spiritual emergence, or religious awakening sometimes feel mentally and emotionally unstable. On the one hand, they may feel totally disconnected from their normal lives with no way of communicating their experience. On the other, they may feel impelled to speak and draw criticism and ridicule. Unfortunately, for some, self-destruction and death can appear to be the only solution. Others may return with a profound spiritual insight but have difficulty holding on to it. Like bringing a dream back to waking consciousness, it can be hard to remember an experience that doesn't fit into your personal or cultural CRO.

On the other hand, if we are able to hold on to profound experiences from other CROs, they can transform our lives. One minute "out of mind" in an ecstatic state can create a renewed sense of purpose that lasts a lifetime.

If we have any positive or negative reaction to *Bleep,* or to the implications of quantum physics, we should pay attention. Our reactions may put us in touch with the underlying assumptions of our CRO. The question may not be "What is real?" but "How could we know any 'true' reality if we're always perceiving through the limiting filters of our personal CRO?" We all are open to—or resistant toward—different perspectives depending on our internal CRO. When we want to believe something, we do, regardless of evidence to the contrary. The thousands of followers of Jim Jones at Jonestown were sure they were on a path to a truer reality. Those of us who weren't "liberated" couldn't possibly comprehend what the dead experienced. We considered ourselves to be the survivors. We were sure they were all wrong.

When we do shift perspectives, we come away from each "mistaken" trance feeling we understand reality better, and we're living in a truer reality. In this way, each CRO shift becomes a new trance, in which the previous one is judged unreal. In terms of absolute reality, neither view is any more real. Shadows are just as real and true when shadows are all we know. It is only the rational mind that seeks to objectify an absolute truth and pass judgment. The creativity of consciousness, however, might mean that we will always be shifting our awareness and perspective. Who's to say our current CRO is the "true" one?

People in the New Age community see *Bleep* as confirmation of the reality they know, or would like to know. For the materialists, the film is obviously mistaken and promotes a dangerous illusion threatening the very basis of reality. Both these conclusions support and arise out of preexisting ideas.

Knowing our minds are very good at fitting the world into the patterns we expect is where a grounding in scientific thinking can be helpful. We can form an idea, observe and experience, then decide if our knowledge and experience fit the idea. Scientific thinking requires that we test our ideas against our actual experience, not our preferences. The problem is, we can't agree on what constitutes valid experience. The realness of an experience shifts in different CROs. Materialistic science says only what can be objectively studied is real; then it falls short when determining the reality of intangibles like gravity, love, and consciousness. Yet scientists have just as much faith in the existence of gravity as the religious have in their god. The spiritual traditions say it's only consciousness, the "I am" inside every experience, that is the ultimate truth. Both views may be correct as observed within the limitations of each CRO.

Quantum physics brings the scientist and the spiritualist closer together. David Bohm, a quantum physicist, became fascinated about how consciousness affects reality. He proposed that our language imposes strong, subtle pressures to see the world as fragmented and static. Thought tends to create fixed structures in the mind, which can make dynamic entities seem static. Bohm would say a noun is just a "slow" verb, that it refers to a process that is progressing so slowly as to appear static. For example, the paper on which this text is printed appears to have a stable existence, even though we know at this very moment it is changing and evolving toward dust.

In quantum physics, just observing quanta—the most basic bundles of energy and information—causes them to collapse into either a physical electron or an energy wave. Similarly, our thought collapses the unrestrained creativity of the universe into tangible objects that are only shadows of their full meaning. We do this every time we think a thought or use language to describe something.

As long as we seek reality only through our logical minds and the language that drives our thoughts, we are inherently limited. Seeking a "true" reality, which is also a thought construct, limits our awareness. Bohm would say there could no more be an absolute true reality than there could be a true sunrise, kiss, or poem. The truth we seek to know may, in fact, be a creative process that cannot be conceptualized. The verb *truthing* would be a better description of something we experience as more real or meaningful.

It's no wonder Bohm's research in quantum physics led him in his later years to study consciousness and meaning. He felt that if we could maintain an awareness of thought processing while in dialogue with people or the physical world, we could learn to suspend our implicit assumptions and beliefs. Bohmian dialogue is aimed not at achieving a particular truth or convincing another of your view, but rather at sharing an experience of meaning. I believe this is partly why *Bleep* has had such a strong impact. It attempts to open this kind of dialogue and draws us out of our dominant CRO. Any shifting of consciousness is meaningful, regardless what truth or untruth the mind wants to put upon it.

Tart, who has researched consensus trance and hypnosis extensively, notes how individuals in deep hypnosis, as in other ecstatic trance states, are able to shift beyond their CRO. I have also seen this in my work as a hypnotherapist. If you repeatedly ask a person going into trance, "Who are you?" the answer changes as he or she transitions into the altered state. Initially one may respond with name, job, or another label. In the profound trance, these parts drop away and he experiences himself less confined to any particular personality or structure. In an article on deep hypnosis, Tart described how one client became more and more identified with what seemed to be ultimate potential. He felt that he could evolve into anything, literally, without limits. His experience echoes the innate potentiality of the quantum realm.

Critics of quantum physics claim the theories don't apply to the reality we live in. Of course, we'll never see quantum effects when we're focused exclusively on objective, physical reality. Since profound inner experiences are not valid, they are ignored. But mystics, people in profound trances, and others who've escaped the CRO have had experiences that reflect quantum possibilities. These people also claim the importance of nonrational states in determining what is real for them.

I know through my own experience and observations of my clients how shifting of CRO is meaningful. Changing perspective brings a new level of awareness back into our lives. With each small shift we make—a near-death experience isn't necessary or desirable—we become less attached to one particular view and more open in our approach to life.

I appreciated how *Bleep* asked us to imagine creatively, to expand out of our CRO. As Einstein said, "I did not discover relativity by rational thinking alone." Science must include creativity and openness in order to access knowledge. Other perspectives must be experienced to know their value. If you've known only objective rationality, you will neither find meaning in nonrational, immersed experiences nor will you respect them.

The film does offer one practical exercise to experience a different worldview. Dr. Joe Dispenza says he creates his day this way: "I wake up in the morning, and I consciously create my day the way I want it to happen." He spends a few moments envisioning himself living as a genius. Then he gets on with his day and waits for a response. "During parts of the day, I'll have thoughts that are so amazing, that cause a chill in my physical body, that have come from nowhere." These thoughts and feelings affirm his intentions and, more importantly, give him the experience of creating his reality.

In the self-hypnosis classes I teach, I have seen similar shifts in awareness. A student will create an effective autosuggestion, then go into trance and repeat the suggestion to his subconscious mind. Then, as he goes about his life, he experiences meaningful shifts relating to his intentions. Obviously, none of this is proof of consciousness affecting reality, but people feel more in control of their lives and they begin living and behaving differently. The effects are real. If you practice setting an intention, you will also experience a shift in your perspective. This is one of the main reasons I also practice and teach remote viewing. The experience of directly perceiving something at a distance forces me to shift out of the CRO of separation and materialism.

The question is: Are you in a place where you want to stabilize your CRO or are you interested in trying on another view? If you want to experience a change, see *Bleep* for yourself. It's now out on DVD. Join the discussions that surround it. But be aware, if you want to know the reality that quantum physics proposes, you have to get actual experience. Discussions aren't enough. You must seek out situations forcing you to expand and perceive other viewpoints. This can be uncomfortable, but is well worth the effort.

I'll help you with a simple, practical exercise for shifting CRO. You'll need a radiometer, a light source, and an open mind. A radiometer is a scientific instrument that looks like a lightbulb with a movable vane suspended inside a near vacuum. Light hitting the surface of the vane causes it to spin. Set the radiometer on a flat, stable surface under a light source and enjoy the whirling movement. Without interfering with the light, you can stop the vane with your mind. When I do this exercise, I start by focusing on the movement of the spinning vane. I imagine myself merging with it and feel the movement in my body. When I sense a connection, I quiet my mind, calm my body, and the vane starts slowing down with me. (Knowing how to meditate or do self-hypnosis is helpful.) When it stops, I see before me proof of my intention. Try it yourself. You might even find another way that suits you better.

Being able to stop the radiometer may have little practical value in your life. You will know from the experience, however, that your mind is able to influence physical reality. This knowledge has tremendous value, and may shift your bleeping reality!

42 From Apollo to Zero Point

When Is a Walk on the Moon
Not the Highest Point in Life?

J. Douglas Kenyon

Growing up on the family ranch near Roswell, New Mexico, in the 1940s provided Apollo 14 astronaut and paranormal researcher-in-the-making Edgar Mitchell with more than a few clues to his destiny. On the way to school, for example, he would walk past the house of reclusive rocket scientist Robert Goddard, whose obscure experiments in the 1920s had inspired the German ballistic missiles of World War II and paved the way for Mitchell's own lunar mission, yet a quarter century away. There were also aircraft of the wood and cloth variety available for flying—an opportunity not lost on the young test pilot to be (his first solo flight came at fourteen). As a youth, Mitchell watched and wondered at the mysterious glows that filled the night skies over nearby White Sands as an atomic age

Fig. 42.1. Astronaut Edgar Mitchell. (Photograph courtesy of NASA)

was being hatched in secrecy. And later another, perhaps stranger, episode, the purported crash of a flying saucer just a few miles away, would also leave intriguing clues to be pondered in a future still—a half century later—in the process of unfolding.

One of the few humans known to have viewed the earth as an "extra-terrestrial," and one of only twelve—so far as we know—to have actually set foot on another celestial body, Mitchell, with cowriter Dwight Williams, has just finished a new book, *The Way of the Explorer* (New York: Putnam, 1996), relating the many experiences in space and on earth that render the universe a far more marvelous and mysterious place than the titans of established science—and, for that matter, most of his fellow astronauts—have dared to admit.

In the book, Mitchell details his widely and sensationally publicized—yet fully scientific—attempt to communicate telepathically from the moon with colleagues back on earth, and goes on to describe the experiment's virtually unreported "dramatic" and positive results. But it was on the trip back to earth during that 1971 mission that he made his most significant encounter with infinity—an experience that was to change his life forever and lead to some of the revolutionary, albeit controversial, conclusions in his book.

He writes, ". . . as I looked beyond the earth itself to the magnificence of the larger scene, there was a startling recognition that the nature of the universe was not as I had been taught. My understanding of the separate distinctness and the relative independence of movement of those cosmic bodies was shattered. There was an upwelling of fresh insight coupled with a feeling of ubiquitous harmony—a sense of interconnectedness with the celestial bodies surrounding our spacecraft."

For Mitchell, the experience, which he would later describe as an epiphany, was so profound and moving that he knew his life had changed irreversibly. Though he continued briefly with the space program and served on the backup crew for Apollo 16, he soon went on to establish, in the early seventies—for the purpose of investigating many of the

Fig. 42.2. Edgar Mitchell walks on the moon. (Photograph courtesy of NASA)

*Fig. 42.3.
Astronaut Edgar
Mitchell returns
to Earth after his
successful moon-
walking mission.
(Photograph
courtesy of NASA)*

questions that had come to preoccupy him—the Institute of Noetic Sciences.

Having earned a doctorate in aeronautics and astronautics from MIT, Mitchell was acutely aware of the failures of Western science to deal with the perplexing problems of consciousness and nonphysical reality. His own observations had already provided plenty of data that failed to square with prevailing views of the possible.

Soon Mitchell encountered Norbu Chen, an American trained in Tibetan Buddhism who, to his amazement, successfully healed his mother of chronic eye problems and thereafter provided plenty of material for investigation. Later he met Uri Geller (the Israeli psychic who was to become famous for his spoon bending abilities), and subsequently sponsored numerous experiments to establish the truth of what was happening. (Mitchell insists that Geller has not been successfully debunked—as has been claimed—and that it is, in fact, the debunkers who have some explaining to do.)

His own research, plus results from some of the more exotic experiments on the frontiers of science, have led Mitchell, in an effort to account for evidence of "the non-local interconnectedness of things," to offer in his book what he calls a "dyadic" model to explain things. The universe, he concludes, is formed of inseparable pairs called dyads, which emerge into time and space from a "zero point"—the intelligent self-generating source of the universe, where all information is stored and never lost, and with which it is possible to resonate and thus, theoretically, to gain access to all knowledge—another way of describing what some religions term enlightenment.

Zero point he defined recently as "[having] zero dimensions, as in mathematics, a point, not a line, plane, or solid—just quantum fluctuation— working like a mirror to create a virtual image, which is building up resonance." Fascinated by the efforts of Nikola Tesla, John Keeley, and others who have attempted—with apparent success—to tap a universally available source of energy, Mitchell sees possible corroboration for his ideas. "If they are correct and many people think they are," he says cautiously, "[their power source] probably is what we call a zero-point field, with non-local interconnected properties."

One experiment in particular played a key role in his thinking. A physicist at the University of Paris named Alain Aspect demonstrated that subatomic particles originating from the same source, though separated by great distances, still managed to maintain the proper quantum relationship to each other, despite any changes that might occur to one or the other. The implication was that communication of some sort is occurring between particles over great distances not limited by the speed of light.

Recently Mitchell agreed to share his thoughts with me. I reached him at his home in Florida, where he lives with his third wife, Sheila, and his teenage son Adam. After quieting one of his schnauzers and settling down with a cup of herbal tea, the former space explorer talked about his book, his theories, UFOs, government cover-ups, ancient mysteries, and other controversies.

The Aspect experiment notwithstanding, communication with fellow astronauts has been limited over the years, though occasionally he does talk to some of them, depending on the subject. "Many of the people in my business, after my flight," he says with a chuckle, "came into my office and said 'Tell me about what you are doing, it's exciting' but they looked furtively as they came in and closed the door very carefully."

Closed doors are nothing new to Edgar Mitchell when it comes to finding mainstream acceptance for his ideas, but he is reluctant to criticize. Though admitting that there is resistance in some quarters, he prefers to make the point that verifiable proof in this area is hard to come by: "We're dealing with levels of nature that are exceedingly subtle, and require a great deal of sophistication in testing them and a lot of money." If there is a problem, he prefers to say, it is with the peer review system in which professional journals decide what is and is not worthy of publication. In that area he's quite willing to say that the system is "atrocious. . . . Too many of the editors, frankly, don't have the skills to be good judges and so they pass these things off. If they don't like it they pass it off to somebody they don't think will like it. If they do like it, they'll pass it to someone they think will like it. The peer review process is just terribly political." Again, tempering his words, he insists that he has no objection to the process in theory.

The difficulty, as with most areas of human function, is hypocrisy: "We talk about the beauty of science, the objectivity, but we let our emotions, our power plays, our greed, and so on—our human fallibility—get into virtually everything we do, including the peer review process." Will he stipulate that in many cases individuals are more concerned with preserving their own prerogatives than the truth? "Absolutely!" The realm of ideas has evolved like most other kinds of politics, he says. "We have stopped burning witches at the stake, but we have certainly not stopped persecuting."

Regarding the role of government in blocking the dissemination of information, however, he is quite willing to cry cover-up, particularly when it comes to questions regarding the famous "crashed UFO" incident at Roswell in 1948. "All you have to do is ask for some information under the Freedom of Information Act," he complains, "and then get in return blacked-out pages, to perceive that. In other words, if you want to know more about it than the notion that it was simply a weather balloon at Roswell, and you ask for answers pertaining simply to that, you get back nothing but the standard pat old answers, filled through and through with censorship that is totally inappropriate to the issue."

Mitchell says he was seventeen at the time of the Roswell incident and didn't personally know any of the principals, though his parents did. In recent years, however, he has been in touch with many who appear to have been very close to the source, including Jesse Marcel Jr. It is clear to Mitchell that many are still frightened about giving testimony. Making no claims of firsthand knowledge in the case, he simply asks "that the people who do have firsthand experience be released from any security oaths and be assured that they won't be prosecuted and that any information relating to the existence of foreign visitors be released." He's optimistic that someday that will happen.

On NBC's *Dateline* in 1996 Mitchell said that he had "met with people from three countries who in the course of their official duties claim to have had close encounters [of the third kind]." On the show he scoffed at the standard Air Force explanation of Roswell as a crashed weather balloon. "The people that were there say that's utter nonsense." Did he think it likely that extraterrestrials have been to this planet? "From what I now understand and have experienced, I think the evidence is very strong, and large portions of it are classified [by the government]." He also told *Dateline* that his information from former highly placed U.S. officials is that the government has picked up engineering secrets from UFOs. *Dateline* was unable to obtain any official response beyond the standard handouts on the subject, stating that "there has been no evidence indicating that sightings categorized as 'unidentified' are extra-terrestrial."

As for the notion that modern scientific knowledge is but the rediscovery of lost ancient knowledge, Mitchell thinks it's only partly true. "What modern science has produced is specificity and a new way of looking at detail and measuring details that the ancients couldn't. They kind of intuitively sensed the broad scale of things. The detail they couldn't know. Putting it together takes science," he says.

On evidence of advanced scientific knowledge by the ancients, such as the engineering and precise alignment of ancient monuments and the astronomical knowledge implicit in their understanding of such phenomena as the precession of equinoxes, Mitchell seems—not surprisingly—to lean toward the ancient astronaut explanation. Fascinated by the work of Zecharia Sitchin, Mitchell would like to see some serious efforts made to validate theories that civilization on earth owes its origins to implantation by extraterrestrials.

Questions on a related topic, though, touch a sore spot. Space researcher and author Richard Hoagland's recent charges in a Washington, D.C., press conference—that the astronauts of Apollo 12 and 14 were actually in the midst of ancient ruins on the moon and that photos were systematically doctored to cover up the evidence—provoke nothing but scorn from Mitchell. The entire event was televised live to the world (making such manipulation virtually impossible), he points out, and adds that Hoagland failed to call him for any kind of comment or corroboration (though Mitchell says he could easily have done so). "I would have given [Hoagland] credit for being persistent and hanging by his guns for saying 'hey, let's look, there's something there worth looking at.' But if he's going to say that it happened on my flight and there's something we missed, or something we're covering up, then he just shot himself in the foot, because we didn't cover up, we didn't miss it. There wasn't anything there. It's just baloney." Mitchell, however, is willing to concede that there may be something to Hoagland's "face-on-Mars" conjectures, as detailed in his book *The Monuments of Mars*. Statistical analysis, Mitchell feels, argues against a purely natural formation on the Cydonia plain. He has long supported a mission to Mars to fully answer such questions.

Whatever he might anticipate from future interplanetary exploration, Mitchell is less sure what to expect from the "undiscovered country" that lies beyond the frontier called death. Though he thinks some kind of survival of identity occurs, he suspects "the mechanism is quite different than we're used to thinking." In Mitchell's view, the accumulated knowledge and experience of an individual—he prefers to describe it as information—remains intact in a universal zero-point field where it can be accessed by other individuals with the appropriate resonance, which, he believes, accounts for data cited in support of reincarnation. In Mitchell's mind there is little difference between such a phenomenon and the classic notion of the soul, though he stops short of believing that discarnate existence outside the three-dimensional world can occur—the software requires the hardware. "Right now a human being is a self aware organism," he explains, "and everything before this instant—right now—is memory. It's just information in your memory or perhaps even somewhere else. What we're proposing here is that the experience—in the form of information—is simply not lost. So in principle, anyone that could claim that information—that total information—is essentially that person."

For Mitchell, zero point is essentially equivalent with God—intelligent, self-organizing, and utilizing information to evolve. "If we in the universe are self-organizing and intelligent and are a product of the universe, then the universe is self-organizing and intelligent and that is also what we ascribe deity to be."

The future for Edgar Mitchell promises to be "more of the same." Which means more books and research into the vast potentials of consciousness, in conjunction perhaps with state-of-the-art media production.

Mitchell has entered into partnership with Hollywood producer Robert Watts (credits include all of the major Lucas and Spielberg movies including the "Star Wars" and "Indiana Jones" series) and others to form North Tower Films. The goal is to create the kind of consciousness-raising material that can help catalyze the needed changes on our small planet. Mitchell thinks the media can play a dominant role in such a process "as much as scientists," but, he points out, "the media [have] to go back to objective reporting."

A world in which science, government, and the media perform their role without bias: It sounds like a star to aim for. For Edgar Mitchell, it's already clear, the moon was just a stepping-stone to infinity, both without and within. Hopefully, the rest of humanity will soon get the opportunity also to make such discoveries, without interference from civilization's established institutions. If not, those institutions may find themselves as outdated as aircraft made from wood and cloth.

Selected Bibliography

Chapter 1. Debunking the Debunkers

Moody, Dr. Raymond. *Life After Life: The Investigation of the Near-Death Experience*. New York: HarperOne, 2001.

Ring, Dr. Kenneth. *Life at Death: A Scientific Investigation of the Near-Death Experience*. New York: William Morrow & Co., 1982.

Chapter 2. "Voodoo Science" on Trial

Park, Robert L. *Voodoo Science: The Road from Foolishness to Fraud*. Oxford: Oxford University Press, 2000.

Rosenblum, Art. "An Interview with Dr. Randell L. Mills of BlackLight Power, Inc." *Infinite Energy* 17 (Dec. 97/Jan. 98): 21–35.

Zubrin, Robert. *The Case for Mars*. New York: Touchstone, 1996.

Chapter 4. Inquisition—The Trial of Immanuel Velikovsky

Velikovsky, Immanuel. *Worlds in Collision*. New York: Dell, 1977.

Chapter 5. The High Technology of the Ancients

Radka, Larry Brian. *The Electric Mirror of the Pharos Lighthouse and Other Ancient Lighting*. Parkersburg, WV: Einhorn Press, 2006.

Chapter 6. A Scientist Looks at the Great Pyramid

Fix, William. *Pyramid Odyssey*. New York: Smithmark Publishing, 1978.

Chapter 7. Precession Paradox: Was Newton Wrong?

De Santillana, Giorgio, and Hertha von Dechend. *Hamlet's Mill: An Essay Investigating the Origins of Human Knowledge and Its Transmission Through Myth*. Cambridge, MA: Harvard University Press, 1969.

Chapter 8. The Dogon as Physicists

Temple, Robert K. G. *The Sirius Mystery: New Scientific Evidence of Alien Contact 5,000 Years Ago*. Rochester, VT: Destiny Books, 1998.

Chapter 9. The Astronomers of Nabta Playa

Bauval, Robert, and Adrian Gilbert. *The Orion Mystery*. New York: Three Rivers Press, 1995.

Brophy, Thomas. *The Origin Map: Discovery of a Prehistoric, Megalithic, Astrophysical Map and Sculpture of the Universe.* Lincoln, NE: Writers Club Press, 2002.

Wendorf, Fred. *Holocene Settlement of the Egyptian Sahara.* New York: Kluwer Academic/Plenum Publishers, 2001.

Chapter 10. Tesla, a Man for Three Centuries
www.aetherometry.com

Chapter 11. Tom Bearden Fights for Revolutionary Science
Bearden, Thomas. *Excalibur Briefing.* N.l.: Strawberry Hill Press, 1980.

Chapter 12. Sonofusion
To view the article "Evidence for Nuclear Emissions During Acoustic Cavitation," by R. P. Taleyarkhan and C. D. West online, go to www.sciencemag.org and do a search on the article's title.

To view the article "Skepticism Greets Claim of Bubble Fusion," by R. P. Taleyarkhan online, go to www.PhysicsToday.org and do a search on the article's title.

Chapter 13. Escape from Gravity
Clarke, Arthur C. *The Final Odyssey.* New York: Ballantine, 1997.
www.projectearth.com

Chapter 14. Power from the Nightside
Verne, Jules. *Journey to the Center of the Earth.* New York: Penguin, 1965.

Chapter 15. Techno Invisibility
Ufimtsev, Pyotr. "Method of edge waves in the Physical Theory of Diffraction." Air Force System Command, Foreign Tech. Div. Document ID No. FTD-HC-23-259-71 (1971).

Vaupel, Elisabeth. *Angewandte Chemie International* 44, Issue 22: 3344–3355.
www.aviation.ru/okb.php
www.ee.duke.edu/~drsmith/about_metamaterials.com
www.ee.duke.edu/~drsmith/cloaking.html
www.sciencemag.org

Chapter 16. Weather Wars
Begich, Dr. Nick, and Jeane Manning. *Angels Don't Play This HAARP: Advances in Tesla Technology.* Anchorage: Earthpulse Press, 1995.
www.cheniere.org/misc/brightskies.htm

www.mosnews.com/news/2005/09/08/kgbkatrina.shtml
www.nexusmagazine.com
www.rense.com/general18/mn.htm

Chapter 19. Madame Curie and the Spirits
Blum, Deborah. *Ghost Hunters: William James and the Search for Scientific Proof After Death*. New York: Penguin, 2007.

Chapter 20. India's Mystic Military
Hatcher-Childress, David. *Antigravity and the World Grid*. Kempton, IL: Adventures Unlimited Press, 1987.
www.indiadaily.com/editorial/2251.asp

Chapter 21. Is the Big Bang Dead?
Arp, Halton. *Quasars, Redshifts and Controversies*. Berkeley: Interstellar Media, 1987.

Chapter 22. The Cycles of Danger
Benton, Michael J. *When Life Nearly Died: The Greatest Mass Extinction of All Time*. London: Thames & Hudson, 2005.
Hoyle, Fred, and Chandra Wrickramasinghe. *Diseases from Space*. New York: Harper & Row, 1980.

Chapter 23. Healing Vibes
Gerber, Dr. Richard. *Vibrational Medicine: The #1 Handbook of Subtle-Energy Therapies*. Rochester, VT: Bear & Co., 2001.
www.soundstrue.com

Chapter 24. The Malady in Heart Medicine
McGee, Dr. Charles T. *Healing Energies of Heat and Light*. Coeur d'Alene, ID: Medipress, 2000.
———. *Heart Frauds: Uncovering the Biggest Health Scam in History*. Colorado Springs: Piccadilly Books, 2001.

Chapter 25. Energy Medicine in the Operating Room
Motz, Julie. *Hands of Life: Using Your Body's Own Energy Medicine for Healing, Recovery, and Transformation*. New York: Bantam, 2000.

Chapter 26. Getting Left and Right Brains Together
Shlain, Leonard. *The Alphabet Versus the Goddess: The Conflict Between Word and Image*. New York: Viking Penguin, 1999.
———. *Art and Physics: Parallel Visions in Space, Time, and Light*. New York: HarperPerennial, 2007.

————. *Sex, Time, and Power: How Women's Sexuality Shaped Human Evolution*. New York: Viking, 2003.

Chapter 27. X-Ray Vision and Far Beyond
www.baytoday.ca
www.baytoday.ca/content/news/details.asp?c=6657
www.baytoday.ca/content/news/details.asp?c=8267

Chapter 28. The Biology of Transcendence
Bramley, William. *The Gods of Eden*. New York: Avon, 1993.
Clarke, Arthur C. *Childhood's End*. New York: Ballantine, 2001.
Kelleher, "Retrotransposons as Engines of Human Bodily Transformation." *Journal of Scientific Exploration* 123, no. 1 (Spring 1999): 9–24.
www.trufax.org

Chapter 29. Paranormal Paratrooper
Morehouse, David. *Psychic Warrior*. New York: St. Martins Press, 1998.
Targ, Russell, and Harold E. Puthoff. *Mind Reach*. New York: Dell, 1978.

Chapter 30. Psychic Discoveries since the Cold War
Ostrander, Sheila, and Lynn Schroeder. *Psychic Discoveries Behind the Iron Curtain*. Upper Saddle River, NJ: Prentice Hall, 1971.
————. *Supermemory: The Revolution*. New York: Carroll and Graf, 1991.

Chapter 31. When Science Meets the Psychics
Hibbard, Whitney S., Raymond Worring, and Richard Brennan. *Psychic Criminology*. Springfield, IL: Charles C. Thomas Publishers, 2002.

Chapter 32. Rupert Sheldrake's Seven Senses
Sheldrake, Rupert. *Dogs That Know When Their Owners Are Coming Home, and Other Unexplained Powers of Animals*. New York: Three Rivers Press, 1999.
————. *The Sense of Being Stared At*. New York: Crown Publishers, 2003.
————. *Seven Experiments That Could Change the World*. Rochester, VT: Park Street Pres, 2002.
www.sheldrake.org

Chapter 33. Telephone Telepathy
Sheldrake, Rupert, and Pamela Smart. "Testing for Telepathy in Connection with E-mails." *Perceptual and Motor Skills* 101, 771–86.
www.abc.net.au/rn/inconversation/stories/2006/1754367.htm#

www.bbc.co.uk/radio4/science/thematerialworld_20060907.shtml
www.earthboppin.net/talkshop/feelers
www.sheldrake.org/Articles&Papers/papers/telepathy/index.html
www.sheldrake.org/D&C/controversies/Times_editorial.html
www.sheldrake.org/D&C/controversies/Times_report.html

Chapter 35. The Fight for Alien Technology
To view the article "New Face of the Mafia in Sicily. High-tech Transformation—
    With Global Tentacles" online, go to AmericanMafia.com and do a
    search on the title.
www.aliensci.com
www.roswellinternet.com

Chapter 36. Altered States
Tart, Charles. *Altered States of Consciousness*. New York: Harper, 1990.

Chapter 37. Searching for the Unifying Field
McTaggart, Lynne. *The Field: The Quest for the Secret Force of the Universe*. New York: Harper, 2003.
———. *What Doctors Don't Tell You*. New York: HarperCollins, 2005.

Chapter 38. The Mind-Matter Connection
Radin, Dean. "Exploring Relationships Between Physical Events and Mass
    Human Attention: Asking for Whom the Bell Tolls." *Journal of Scientific Explanation* 16, no. 4 (Winter 2003): 533–47.

Chapter 39. Dr. Quantum's Big Ideas
Pert, Dr. Candace. *Molecules of Emotion: The Science Behind Mind-Body Medicine*. New York: Simon & Schuster, 1999.
Wolf, Fred Alan. *Dr. Quantum Presents: A User's Guide to Your Universe*
    [AudioBook] (Audio CD). Louisville, CO: Sounds True, 2005.
———. *Taking the Quantum Leap: The New Physics for Non-Scientists*.
    New York: HarperPerennial, 1989.

Chapter 40. Further Explorations of the Crack in the Cosmic Egg
Pearce, Joseph Chilton. *The Biology of Transcendence: A Blueprint of the Human Spirit*. Rochester, VT: Park Street Press, 2004.
———. *The Crack in the Cosmic Egg*. Rochester, VT: Park Street Press, 2002.
———. *Evolution's End: Claiming the Potential of Our Intelligence*. New York: HarperOne, 1993.

————. *Magical Child*. New York: Plume, 1992.
www.ttfuture.org

Chapter 42. From Apollo to Zero Point
Mitchell, Edgar, and Dwight Williams. *The Way of the Explorer*. New York: Putnam, 1996.

Contributors

Amy Acheson, deceased, was a freelance journalist and researcher who studied planetary catastrophism for forty years. For several years before her death she and her husband, Mel Acheson, collaborated with Wallace Thornhill and Dave Talbott on the Internet newsletter THOTH, which features the convergent theories of Thornhill and Talbott.

Peter Bros disagreed at an early age with the accepted explanation that objects fall because it is a property of them to fall and he went on to challenge the current, splintered concepts of empirical science by taking an advanced science curriculum at Bullis Preparatory School, a degree in English at Maryland University, and a doctor of jurisprudence at Georgetown. The result is *The Copernican Series,* a multivolume exposition that sets forth a consistent picture of physical reality and humanity's place in the universe.

John Chambers has a M.A. in English from the University of Toronto and spent three years at the University of Paris. His translations include "Phase One: C. E. Q. Manifesto," in *Quebec: Only the Beginning.* He has published numerous articles on subjects ranging from ocean shipping to mall sprawl to alien abduction, and is the author of *Victor Hugo's Conversations with the Spirit World* (1998 and 2008). Seven of his essays appeared in *Forbidden Religion: Suppressed Heresies of the West* (Bear & Company, Rochester, Vermont). He is the director of New Paradigm Books Publishing Company (www.newpara.com).

Walter Cruttenden is the director of the Binary Research Institute, an archaeoastronomy think tank located in Newport Beach, California. His focus is on the astronomy, mythology, and artifacts of ancient cultures, with an emphasis on history theory and cycles of consciousness. He is the author of *Lost Star of Myth and Time* (St. Lynn's Press), a studied look at ancient cultures throughout the world and their belief in a vast cycle of time. Previously Cruttenden wrote and produced the award-winning PBS documentary *The Great Year,* narrated by James Earl Jones, which looks at the myth and folklore of ancient cultures and seeks to find the message that these cultures left for modern humanity.

William P. Eigles is a director of the International Remote Viewing Association, which promotes research and education concerning scientifically validated paranormal perception, and the managing editor of its quarterly publication, *Aperture*. Educated as a biomedical engineer in Canada and an attorney in Colorado, he worked in the computer and telecommunications industries for fourteen years. He currently serves as a noetic adviser, using astrology, remote viewing, transpersonal hypnosis, and other intuitive skills to assist people on their life paths to understand and predict world events.

Mark H. Gaffney is a researcher, writer, poet, environmentalist, peace activist, and organic gardener. He was the principal organizer of the the first Earth Day at Colorado State University, in April 1970. Gaffney's first book, *Dimona: the Third Temple?* (1989), was a pioneering study of the Israeli nuclear weapons program. His latest book, *Gnostic Secrets of the Naassenes,* was a finalist for the 2004 Narcissus Book Award. The book has since been translated into Greek and Portuguese. He is currently writing a book about the events of September 11, 2001. Feel free to visit his website at www.gnosticsecrets.com.

William Hamilton III is a senior software developer who has worked in the information technology field for over thirty years. He was employed from 1961 to 1965 by the U.S. Air Force Security Services, where he received the top grade in electronic technology. A psychology major at Cal State Universtity in Los Angeles, he received an A.A. degree in physical science at Pierce College in 1987 and also studied information technology at the University of Phoenix. Past and present affiliations include the Foundation for Research in Parapsychology (1960–1961), Understanding, Inc. (1957–1961), the World Federation of Science and Engineering (1970s), MUFON (1976 to present), and Skywatch International (1997 to present). Membership in high IQ societies include MENSA (1983), GLIA (2002), and IHIQS (2005).

Frank Joseph is a prolific author whose books, published by Bear & Company, Rochester, Vermont, include *The Destruction of Atlantis, Survivors of Atlantis,* and *The Lost Treasure of King Juba.* He has been the editor in chief of *Ancient American* magazine since 1993, is a member of Ohio's

Midwest Epigraphic Society, and was inducted into Japan's Savant Society in 2000. He lives in Colfax, Wisconsin.

Len Kasten is a freelance writer, journalist, and researcher whose articles have appeared in several alternative publications. A graduate of Cornell University, he is the president of the American Philosopher Society, and has been involved in New Age activities for more than twenty-five years. He is the former editor of *Horizons* magazine, and has written over forty published articles for *Atlantis Rising*. He currently resides in Arizona.

J. Douglas Kenyon has spent the last forty years breaking down barriers to paradigm-challenging ideas. Utilizing the media in their various forms, he has consistently pushed points of view largely ignored by the mainstream press. He founded *Atlantis Rising* magazine in 1994, and it has since become a "magazine of record" for ancient mysteries, alternative science, and unexplained anomalies. He is the editor of *Forbidden History* and *Forbidden Religion* (Bear & Company, Rochester, Vermont), thought-provoking books that feature the works of groundbreaking researchers (Graham Hancock, John Anthony West, Zecharia Sitchin, et al.) and challenge the prevailing status quo.

John Kettler is a regular contributor to *Atlantis Rising* on cutting-edge topics. His work has appeared on film in the Academy Award–winning documentary *The Panama Deception* and he is a prolific writer who can also be heard on radio interviews and via Web casts. Formerly a military analyst for Hughes and Rockwell, he presently lives in Woodland Hills, California, and pursues a variety of entrepreneurial, consulting, and editorial projects.

David Samuel Lewis is a journalist who specializes in alternative scholarship dealing with the origins of life, civilization, and human existence. He publishes *The Montana Pioneer*, a monthly news-and-human-interest journal distributed in southwest Montana. He has regularly contributed articles to *Atlantis Rising* that deal with alternative theories of history, science, human origins, and consciousness. Born and raised near Philadelphia, he now makes his home in Livingston, Montana.

Cynthia Logan is a freelance writer specializing in interview profiles with leading professionals in the health, scientific, spiritual, and art communities. Striving to bridge new ideas with mainstream thought, her work has been featured in *Atlantis Rising* magazine since its inception. She lives in Bozeman, Montana, where she regularly contributes to several regional magazines and is the managing editor of *The BoZone,* a bimonthly arts/ entertainment publication.

Eugene Mallove, deceased, was editor in chief of the bimonthly magazine *Infinite Energy* (founded in 1995) and president of the nonprofit New Energy Foundation (founded in 2003). He received a bachelor of science degree and a master of science degree (in aeronautical and astronautical engineering) from the Massachusetts Institute of Technology and his Ph.D. in environmental health sciences from Harvard University. He had broad experience in high-technology engineering. From 1987 to 1991 he was chief science writer at the MIT News Office. He authored three books: *Fire from Ice: Searching for the Truth Behind the Cold Fusion Furor* (1991), *The Starflight Handbook: A Pioneer's Guide to Interstellar Flight* (1989), and *The Quickening Universe: Cosmic Evolution and Human Destiny* (1987). Dr. Mallove was the technical adviser to the 1997 movie *The Saint* and writer and producer of the 1999 documentary *Cold Fusion: Fire from Water.*

Jeane Manning is a sociologist and longtime researcher of quantum-leap energy systems that could replace oil—and the implications these systems have for humankind. She authored *The Coming Energy Revolution* and coauthored nonfiction books including *Angels Don't Play This HAARP,* with Dr. Nick Begich. These books are published in six languages and she has been featured speaker at several energy conferences in Europe. As an *Atlantis Rising* columnist, she reports on breakthroughs bimonthly.

Patrick Marsolek is the director of Inner Workings Resources, researching archaeology, altered states of consciousness, hypnosis, intuition, and dialogue. He is a clinical hypnotherapist and has authored *Transform Yourself: a Self-Hypnosis Manual* and a series of self-hypnosis and relaxation CDs. He teaches and lectures on self-empowerment, self-actualization, and developing extended capacities; he also leads experiential intuitive field trips to sacred sites. See www.innerworkingsresources.com for more information or to contact him.

Susan B. Martinez, Ph.D. (anthropology, Columbia University), is a freelance writer and independent scholar, specializing in the "New Science" of *Oahspe,* as derived from its "Book of Cosmogony and Prophecy." She is author of *The Psychic Life of Abraham Lincoln* and also serves as book review editor at the Academy of Spirituality and Paranormal Studies. As a spiritualist, she has written articles on the Overshadowing, and is currently myth-masher extraordinaire, writing to disprove the Ice Age, global warming, and reincarnation.

Robert M. Schoch, a full-time faculty member at the College of General Studies at Boston University since 1984, earned his Ph.D. in geology and geophysics at Yale University. Schoch has been quoted extensively in the media for his pioneering research recasting the date of the Great Sphinx of Egypt, as well as for his work on ancient cultures and monuments in such diverse countries as Peru, Bosnia, Egypt, and Japan. Schoch has appeared on many radio and television shows and is featured in the documentary *The Mystery of the Sphinx*. Dr. Schoch is the author or coauthor of both technical and popular books, including the trilogy with R. A. McNally: *Voices of the Rocks* (1999), *Voyages of the Pyramid Builders* (2003), and *Pyramid Quest* (2005). His website is located at www.robertschoch.net.

Laird Scranton is an independent software designer who became interested in Dogon mythology and symbolism in the early 1990s. He has studied ancient myth, language, and cosmology for nearly ten years and has been a lecturer at Colgate University. He is the author of *The Science of the Dogon* and *Sacred Symbols of the Dogon*. He also appears in John Anthony West's *Magical Egypt* DVD series.

About *Atlantis Rising*

Atlantis Rising is a bimonthly magazine that presents breaking news from cutting-edge researchers in the fields of Ancient Mysteries, Future Science, Unexplained Anomalies, Paranormal Research, and Alternative Energies.

To subscribe to *Atlantis Rising*, please write to

> Atlantis Rising
> P.O. Box 441
> Livingston, MT 59027
> 800-228-8381
>
> or visit their website
> www.AtlantisRising.com

BOOKS OF RELATED INTEREST

Forbidden History
Prehistoric Technologies, Extraterrestrial Intervention,
and the Suppressed Origins of Civilization
Edited by J. Douglas Kenyon

Forbidden Religion
Suppressed Heresies of the West
Edited by J. Douglas Kenyon

Infinite Energy Technologies
Tesla, Cold Fusion, Antigravity, and the Future of Sustainability
Edited by Finley Eversole, Ph.D.
Foreword by John L. Petersen

Origins of the Sphinx
Celestial Guardian of Pre-Pharaonic Civilization
by Robert M. Schoch, Ph.D., and Robert Bauval

Black Genesis
The Prehistoric Origins of Ancient Egypt
by Robert Bauval and Thomas Brophy, Ph.D.

Lost Knowledge of the Ancients
A Graham Hancock Reader
Edited by Glenn Kreisberg

Secrets of Antigravity Propulsion
Tesla, UFOs, and Classified Aerospace Technology
by Paul A. LaViolette, Ph.D.

Lost Technologies of Ancient Egypt
Advanced Engineering in the Temples of the Pharaohs
by Christopher Dunn

Inner Traditions • Bear & Company
P.O. Box 388
Rochester, VT 05767
1-800-246-8648
www.InnerTraditions.com

Or contact your local bookseller